ENCYCLOPAEDIA OF
OPTICS

ENCYCLOPAEDIA OF OPTICS

Volume 3

Optics and Light

Mariyam Noor

ANMOL PUBLICATIONS PVT. LTD.
NEW DELHI - 110 002 (INDIA)

ANMOL PUBLICATIONS PVT. LTD.
4374/4B, Ansari Road, Daryaganj
New Delhi - 110 002
Ph.: 23261597, 23278000
Visit us at: www.anmolpublications.com

Encyclopaedia of Optics

First Published, 2006

ISBN 81-261-2613-2 (Set) 3

[Responsibility for the facts stated, opinions expressed, conclusions reached and plagiarism, if any, in this volume is entirely that of thc Author/Editor. The Publishers bear no responsibility for them, whatsoever.]

PRINTED IN INDIA

Published by J.L. Kumar for Anmol Publications Pvt. Ltd., New Delhi - 110 002 and Printed at Mehra Offset Press, Delhi.

Contents

Preface

Science is the sermon of today. In every walk of life, since childhood till old age, science plays an important role in the life of man. Uptill some decades back, studied merely as a subject in schools and colleges, Science has assumed the status of an essential part of life today. With the expansion of literacy among the masses and growing awareness, they have woken up from the slumber of ignorance and have started taking up things more sincerely and thoughtfully. It is one of the reasons that our nation is heading towards new goals day after day.

It is a vast subject to study and talk about. For the purpose of ease, Science has been divided into more than two hundred and fifty branches. Specialisation in any branch can be obtained at different institutions and universities, scattered all over the country and abroad.

Apart from other super-specific branches, Physics, Chemistry, Zoology, Botany, Mathematics and Computronics, etc. are more general and closer to daily human life. From the point of view of the cut-throat competition in educational world, a thorough study of the above-mentioned branches, proves very handy, for the new aspirants, especially and for teachers and tutors, generally.

An urgent need was being felt for such a series of books, encompassing the commonly-studied and discussed branches of Science of equal significance and presented in an easy language, understood by all. With God's grace, our dream came true and we are happy, while introducing the dream-series for the lovers of knowledge.

The series consists of many a title, covering various branches of Science, such as Physics, Chemistry, Zoology, Botany, Geology, Agriculture Science, Computer Science, Mathematics and so on. The series is important and a must not only for students but also for teachers. Explained in simple language and aided with graphics and diagrams, everything, between the two ends—title and back cover—of the series, has been presented in a perfect way.

This set of books is devoted to **Optics**, which comprises three volumes: (1) Physical Optics, (2) Geometrical Optics and (3) Optics and Light.

It is hoped that the series would serve both, the students community and the academics, equally and will prove to be a boon for those, aspiring for newer heights. Nevertheless, any suggestions for further betterment of the endeavour will be highly appreciated and solicited.

Editor

1

Light Velocity

The velocity of light in free space is one of the most important of the universal constants. Its value is so large that it evaded experimental determination by early experiments and until 17th century it was believed that light travelled with an infinite velocity. In 1676 Olaf Roemer (1644-1710) a Danish astronomer, made a systematic study of the eclipses of one of the Jupiter's moons and demonstrated for the first time that light had a definite velocity. We, now, know that it is also the velocity with which the wireless waves, X-rays and cosmic rays travel and hence the importance of the subject. Since Roemer's time a large number of methods for the accurate determination of the velocity of light have been developed some of which we shall now describe.

Foucault's Rotating Mirror Method

In 1850 Foucault measured the velocity of light by a new and more accurate method.

It is a modification of an earlier method by Fizeau and needs much shorter distance, even a big room can serve the purpose. Instead of using a toothed wheel as in Fizeau's experiment, he made use of a rotating mirror.

Light from a strongly illuminated narrow rectangular slit A crossed by a fine wire is converged by an achromatic lens L to a plane mirror M capable of rotation at a high speed about a vertical axis through O. This light beam after reflection at the rotating mirror is brought to focus on the concave mirror C whose centre of curvature lies at O, the axis of rotation of M. On reflection from the concave mirror the rays retrace their original path giving rise to an image coincident with A, whatever the position of M.

To observe the image conveniently it is thrown art S′ by means of a thin semi-silvered parallel glass plate mm placed near A at an angle of 45° to the axis of the system. The image is observed by a micrometer eye-piece fitted-with cross-fires.

Now, suppose the mirror M is rotated, the light reflected from -it sweeps over the concave mirror C for a short time in every revolution which after retracing its path gives rise to an intermittent image at S. When, however, the speed is increased, due to rapid impressions on the retina a continuous image is seen but the brightness of this image as much less than that of the image when the mirror is stationary, As the speed is farther increased so that during the time light takes to go from M to C and back, M has moved through a small angle to the position M', the reflected beam no longer retraces its path to S but follows a slightly different path giving rise to the image at S'. In fact when the speed of rotation is very large and uniform the image is drawn out into a band. The displacement SS' depends upon the velocity of rotation and is constant if the velocity is uniform. By measuring the displacement SS′ (= AA'), the speed of rotation, and certain distances, speed of light can be calculated.

Calculations. Let $AL = a$, $LO = b$ and $OC = d$, x = the displacement SS', n the number of rotations of M per sec. and c the velocity of light.

If t seconds be the time taken by light to travel from M to C and back to M, then $t = \frac{2d}{c}$. The angle turned by the mirror in this time is given by

$$\theta = 2\,\pi n t = 2\pi n \times \frac{2d}{c} = \frac{4\pi n d}{c}. \qquad ...(i)$$

When M is at rest, light returning to L appears to diverge from a point B which is the image of C in M. When M has moved through an angle θ, the returning rays appear to come from B' such that

$$OB = OB' = OC = d.$$

Since the reflected ray is turned through twice the angle through which the mirror turns, BB' must subtend an angle 2θ at O,

i.e., $\angle BOB' = 2\theta$.

Hence $BB' = d \times 2d\theta.$ ...(*ii*)

Now SS' or AA' $(= x)$ is the image os BB' formed by the lens L, hence

$BB' : AA' : : BL : LA$

or $$\frac{BB'}{x} = \frac{b+d}{a}. \text{ or } BB' = \frac{(b+d)\,x}{a}. \qquad ...(iii)$$

If follows from (*ii*) and (*iii*) that

$$2d\theta = \frac{(b+d)\,x}{a}, \therefore \theta = \frac{(b+d)\,x}{2ad} \qquad ...(iv)$$

Equating (*i*) and (*iv*),

$$\frac{4\pi n d}{c} = \frac{(b+d)\,x}{2ad}.$$

$$\therefore \qquad c = \frac{4\pi nd \times 2ad}{(b+d)x} = \frac{8\pi nad^2}{(b+d)x} \qquad ...(1)$$

In Foucault's experiment in place of single mirror *C* five such mirrors each of radius of curvature equal to 4 meters were used so that $d = 20$ meters. The displacement of the image was only 0.70 mm which could be measured by the micrometer eye-piece to within 0.005 mm. He calculated the value of *c* to be 2.98×10^8 meters/sec.

Discussion of the revolving mirror method. The main drawbacks in Foucault's experiment are (*i*) the displacement of the image is too small to be measured accurately and (*ii*) the image is very faint owing to several reflections and refraction.

In order to obtain a large displacement it is necessary to increase the distance between two mirrors, since the speed of rotation is limited by mechanical difficulties. But the increase in the distance results in a further diminution of the brightness of the image far during one rotation of the rotating mirror the reflected beam makes two and, therefore, falls on C for a fraction $\delta/4\pi d$ of a revolution Where δ is the width of C. Consequently, the ratio of the brightness *I* of the image when *M* is rotating to the brightness I_0 of the image when *M* is at rest is equal to $\delta/4\pi d$, that is,

$$\frac{I}{I_0} = \frac{\delta}{4\pi d}.$$

This difficulty was overcome by Michelson who repeated the experiment in 1884 by placing an achromatic lens of large aperture and a long focal length between the rotating mirror *R* and the concave mirror C instead of between the slit *S* and the rotating mirror *R* so that light was incident on C for a greater part of the mirror's revolution.

By this arrangement the distance *d* was increased to 600 meters without any loss in brightness of the image. We find that when the mirror R revolves, the image of the slit S is formed on the circumference of the circle of radius OS.

Let S_1S_2 be the images of S for two extreme positions of the rotating mirror R when the lens L allows the reflected light from R to fall on the opposite edges of the mirror C, then S_1S_2 and the width δ of the mirror C may be taken as object and image heights with respect to L.

$$\because \qquad \frac{\delta}{S_1S_2} = \frac{f}{u-f}$$

where f is the focal length of the lens and $S_1L=S_2L=u$

$$\text{or} \qquad S_1S_2 = \delta\,\frac{u-f}{f}$$

For one revolution of the mirror the light falls on the mirror C for a fraction $\dfrac{S_1S_2}{4\pi r}$, where $OS=r$

$$\text{so that} \qquad \frac{I}{I_0} = \frac{S_1S_2}{4\pi r} = \Upsilon \times \frac{u-f}{f} \qquad \text{...(2)}$$

which is independent of the distance OC $(=d)$. It is obvious that if $\dfrac{u-f}{f}$ is constant the intensity of the image is unchanged.

By this device Michelson obtained a displacement of 133 mm or about 200 times that obtained by Foucault.. The focal length of the lens was about 45'72 meters. The revolving mirror was placed 4.57 meters inside the focus of the lens which was of 20.32 cm aperture. The distance between the slit and the revolving mirror was about 8.52 meters. In a large number of observations the speed of R was 258 revolutions per second. Michelson's result was 2.9985×10 meters/sec.

To further increase the brightness of the image the suggestion of New comb to replace the rotating mirror by a four-sided mirror was adopted.

Velocity of light in stationary matter. The first experiment to find the velocity of light in a transparent medium, much denser than air, was performed in 1850 by Foucault who placed a long pipe of water between rotating mirror R and the concave mirror M_2. He found that with the number of revolutions per second of the mirror the same as when air was used the displacement x of the image was greater than before and so proved (see formula, $c = \dfrac{8\pi nad^2}{(b+d)x}$ *cf*) that the velocity of light in water is less than that in air. This result was of the greatest importance since corpuscular theory demanded a greater velocity for light in a more refracting medium. Michelson later found that the ratio of the velocity of light in air to that in water was 1.330, a result in good agreement with the predictions of the wave theory of light. Hence the experiment provided a crucial test between the two rival theories and the wave theory came to be firmly established.

Michelson's Null Method

During 1924-27, Michelson devised a most accurate method for the determination of the velocity of light. He introduced the principle of null displacement by replacing the plane revolving mirror with an octagonal revolving mirror, the equality of the angles between different faces of which was adjusted by interference to within an error of the order of 0.1" to 0.05'. The optical path was arranged to be of such a length that the time taken by the beam in its outward and inward journeys was the same as taken by a reflecting face to exactly replace the preceding adjacent face. The two stations selected were Mount Wilson for the observing station and Mount San Antonio for the distant station about 35 km away.

Light from an arc lamp A after passing through a slit S is reflected at the face a of the octagonal mirror R capable of rotation about an axis perpendicular to the plane of the paper. By means of two small mirrors M_1 and M_2 it is made to fall on a large concave mirror C_1 of about 18.2 meters radius of curvature and 61 cm aperture. From this it is reflected as a parallel beam to a second concave mirror C_2 situated at the distant station. This brings it to focus on a small concave mirror C_3 which sends back the light to C_2, from where it proceeds to C_1 along the original path. The beam is then reflected to a plane mirror M_3 placed at right angles to M_2 which sends it to M_4 and finally to the face b of the rotating mirror R. From there it is sent into the measuring eye-piece T by means of a totally reflecting prism P.

When the apparatus is set and the revolving mirror is stationary, the image of the source is received on the intersection of the cross-wires of the eye-piece. As soon as the octagonal mirror is set rotating the image disappears. The speed of rotation is gradually increased and so adjusted that the image again fails upon the intersection of the cross-wires of the eye-piece. This occurs when during the time light travels to the distant station and back to the face b_2 it is exactly replaced by the adjoining face. In Michael son's experiment the required speed of R was about 528 revolutions/sec. For greater speeds of R the light may be reflected from successive faces.

Taking the light to be reflected from the next adjacent face, we find that R must have turned through an angle $2\pi/8$ radians, while the light travelled to C_2 and back. If ω be the angular velocity of the octagonal mirror and $2d$ the total distance travelled by the light, then the time taken by the mirror to turn through $2\pi/8$ radians is $2\pi/8\omega$ and this is equal to $2d/c$, where c is the velocity of light.

Hence $$\frac{2\pi}{8\omega} = \frac{2d}{c}$$

or $$c = \frac{8d\omega}{\pi} = \frac{16\,\pi nd}{\pi} = 16nd,$$

$$\ldots\left[\because\ \omega = 2\pi n\right]$$

where n is the number of revolutions/sec.

In general for an r-sided mirror

$$c = 2rnd. \quad ...(3)$$

In this experiment light travelled in each of the to and fro journeys a distance of approximately 35 km which was measured to an estimated error of one part in 11 millions or about 3 mm, the octagonal revolving mirror made 528 revolutions per second and this speed was determined by stroboscopic comparison with a standard electrically maintained tuning fork.

The results of about 1600 individual determinations published in 1926 were remarkably consistent and gave the average value for the velocity of light as

$$c = 299796 \pm 4 \text{ km/sec.}$$

Discussion. It may be clearly understood that it is very difficult to maintain the high speed of rotation of the mirror constant for a long time so that the image is stationary for a short time only. Moreover, because of the high speed the rotating mirror is likely to be broken. To overcome this difficulty Michelson increased the number of faces of the mirror and reduced the speed of its rotation correspondingly, keeping the distance between the two mirrors C_1 and C_2 unchanged.

The method has got certain distinct advantages over the previous methods—(*a*) it being a null method, does not require any measurement of the displacement as in Foucault's method; (*b*) the distance *d* between the two stations is quite large ; (*c*) the image is quite bright so that the position of the image can be located with precision ; and (*d*) the appearance and disappearance of the images quite abrupt and it remains on the cross-wires only for a certain constant speed of the mirror.

Determination of the velocity of light in vacuum. In 1929 Michelson in collaboration with G.F. Pease and F. Pearson planned a determination of the velocity of light in vacuum.

This experiment was not completed till after the death of Michelson in 1931.

An iron pipe 91.25 cm diameter and 1.609 km long was constructed so that it could be evacuated by powerful vacuum pumps to a pressure as low as 0.5 mm of mercury. Light from an arc lamp *A* is focused by *L* on the narrow slit *S*. It is then made to fall on one of the faces of a 32-faced mirror *R* which reflects it into the evacuated pipe through a quartz window *W* set in the wall of the steel pipe. Light is received on a plane mirror *p* and after reflection falls on a concave mirror *C* of diameter 1.016 m and focal length 15.24 meters which renders the beam parallel. This beam of light falls at *a* upon a plane mirror *M* fixed at the otner end of the pipe. The beam then returns to a similar mirror *M′* slightly inclined to the first and after suffering reflections at *b*, *c* and *d* it falls normally at *e* and retraces its path to the concave mirror *C*. Then after reflection at *p* it emerges from the pipe through the window *W* and strikes the adjacent face of the revolving mirror *R* and is observed through the eye-piece *L* after total reflection through a right-angled prism *P*.

The total path of light, to and fro, was about 15.09 km. *R* was rotated by an air blast at such a speed that one face just replaced the next in the time required by the light to return. The calculations are similar to the previous experiment. As a result of 2885 observations, the value of velocity of light in vacuum was determined to be 299774 ± 11 km per sec.

Kerr Cell Method

It is an improvement on Fizcau's mechanical shutter method and was devised by *Karolus* and *Mittelstaedt* in 1925. It is based on the application of Kerr cell between two crossed nicols which provide an *electro-optical shutter* capable of interrupting a beam of light several hundred times more rapidly than can be done with a Fizeau's toothed wheel. The cell *K*, consists of a small glass vessel with two sealed-in metal electrodes and filled with pure *nitrobenzene* or thodichlorobenzene depending upon the frequencies used [14 mega-cycles/sec. in the former case and 28.56 mega-cycles/sec. in the latter case]. When it is placed between two crossed nicols N_1 and N no light will be ordinarily transmitted by the combination but when a high frequency alternating voltage is applied to its electrodes, the dielectric nitrobenzene becomes *doubly refracting* for the polarised light and some light emerges from the analysing nicol N. The application of the field is like providing a space and its absence like the Kerr cell, intervening of a tooth as in Fizeau's cogged wheel experiment. In this method *it is essential that principal planes of the nicols are at an angle* π *4 to the direction of the applied electric field.* By applying an electrical oscillator with radio-frequency voltage variations to electrodes of Kerr cell between crossed nicols a light beam can be interrupted at the rate of several million times per second.

Depicts the arrangement employed for measuring the velocity of light by the Kerr cell method. Light from a source S is focused by a lens L_1 on the Kerr cell K_1 placed between two crossed nicols N_1 and N_2 and on emerging through it, is rendered parallel by the lens L_2. It then falls on a system of mirrors represented by M, so that after reflection back and forth from them it is sent to L_2 and is brought to focus at the centre of an exactly similar Kerr cell K_2. It then passes through the nicol N_3 and then through the lens L_4 to the eye. The total path was 332 meters. If N_2 and N_3 are also crossed no light will reach the eye unless the voltage is applied to the electrodes of K_2.

If a high frequency oscillatory voltage is applied to K_1 the nitrobenzene in it becomes doubly refracting when the voltage rises to maximum and some light is transmitted through N_2. But when the voltage applied drops to zero during its variation from positive maximum value to negative maximum value and *vice versa,* the Kerr cell loses its doubly refracting property. Consequently light waves from N_1 are transmitted through K_1 without any modification and therefore refused transmission by N_2, due to its crossed position with respect to N_1. In this way a beam of light leaving N_2 will be broken up into a series of flashes. Suppose now the same high frequency oscillatory voltage is applied to K_1, K_2 synchronously, that is, the two light shutters are made to open and close together and the distance between K_1 and K_2 is adjusted so that by the time a bright flash of light which has passed through N_2 arrives at K_2, the field drops to zero, K_2 loses its doubly refracting property and thus refuses to transmit light. As explained above in this event light will not be transmitted by N_3 due to its crossed position with respect to N_2 and the eye therefore receives no light exactly in the same manner as in Fizeau's experiment when the light is stopped by the next tooth.

Let the effective distance between the two cells when this occurs be d and if f be the frequency of the voltage variations across the cells, then the voltage will fall from maximum to zero value in $1/4f$ sec. This is the time in which light travels the distance d between the two Kerr cells. Hence the velocity of light is given by

$$c = \frac{d}{1/4f} = 4\,fd. \qquad ...(4)$$

Starting with a given distance between the cells and a gradually increasing frequency a series of maxima and minima will be obtained as in Fizeau's experiment.

It may, however, be noted that in cogged wheels and rotating mirror because of necessarily low frequencies with which flashes of light could be sent, we were required to use large distances. With high frequency interruptions as in this case, a relatively short base line is required and the entire apparatus can be contained in one building so that the atmospheric conditions can be controlled. For instance, with an oscillator of frequency 10^7 times/sec., the base line, *viz.*, the distance between the two cells K_1 and K_2 could be reduced to about 7.5 meters.

As a result of 775 determinations Karolus and Mittelstaedt gave a mean value for the velocity of light=299,778 km/sec. ± 20 km/sec..

These experiments have subsequently been repeated by W.C. Anderson in 1941 with a modified form of the apparatus using a relatively short base line, a single Kerr cell and a photoelectric cell in place of the eye.

Anderson's Method

The diagrammatic sketch of the apparatus used by W.C. Anderson of Harward University N_1 and N_2 are the two nicols in crossed position. In between them there is a Kerr cell K whose plates are connected to a source of high frequency alternating voltage. Light from a mercury arc S on passing through the modulator cell K is converted into light pulses which are split into two beams by the half-silvered mirror M_1 so that the reflected beam traversed the short path to M_2 and back through M_1 to a photocell P (a photo-multiplier tube). The other transmitted one after suffering a number of successive reflections at M_3, M_4, M_5 travelled a longer path upto M_6. Then retraced its course to M_1, where it is reflected to the photo-cell. The photo-electric current produced by the joint action of the two beams is amplified by an amplifier tuned to the frequency of the oscillator which drives the Kerr cell. If the two sets of light pulses arrived at P in phase with each other the photo-cell gives a maximum response and none if they arrived out of phase. If f be the wavelength of the sinusoidal light pulses transmitted by the modulator and c the velocity of light, then $f = cT$, where T^x is the periodic time of the oscillator. To find f the following procedure was adopted.

The mirrors M_4, M_5 and M_6 of the longer path were so arranged that twice the distance $M_4 M_5 M_6$ was nearly 11 l, an integral multiple of the period of radio frequency oscillator. The to and fro length of this segment, *i.e.*, 2 M_4 M_5 M_6 was accurately measured with an invar tape to be 171.864 ± 0.002 m, the exact difference between this distance and 11 l was then determined.

First the segment M_4 M_6 M_5 was cut out of the path so that the mirror M_4 returned light directly to M_3. The path difference was adjusted by a micrometer screw moving M_2 to a position where the photo-cell gave the minimum response, *i.e.*, the two beams were out of phase.

Next M_4 was replaced by M_4' so that light was reflected to plane mirror $_6$ before returning to P and the movable mirror M_2 was shifted to the position M_2' so that the photo-cell again gave minimum response.

Calculations. Let s be the distance from M_4 to M_6 ; x the distance from M_1 to M_4 y from M_1 to M_2, Δs the distance M_4 M_4' and Δy the distance $M_2 M_2$. Then for the first adjustment, we have

$$2x - 2y = 1/2 \qquad ...(i)$$

[since the beams are out of phase]

and for the second adjustment

$$(2x + 2s + 2\Delta s) - (2y + 2\Delta y) = 11\frac{1}{2} l. \qquad ...(ii)$$

Subtracting equation (*i*) from equation (*ii*), we have

$$2(s + \Delta s) - 2\Delta y = 11 l.$$

So that it is not necessary to measure either x or y.

The small shift Δs involved in interchanging M_4, M_4' was measured by micrometer callipers as 2.477 ± .012 cm. Thus each measurement of Δy gave a value for l.

The periodic time T of the oscillator was calculated from

its frequency which was accurately maintained at 19.2 mega cycles/sec. by a crystal control to an accuracy of 1 part in a million and was checked by Anderson against the standard frequencies broadcast from Arlington. The value of c was then calculated by the relation $c = l / T$.

This method has a number of advantages.

1. The difficulty of matching the characteristics of the two cells is avoided.
2. The reduction in the electrostatic capacity enables a higher radio frequency voltage to be developed.
3. The photo-cell can judge the light intensities better than the eye and the received signal can also be amplified.

Anderson made 2895 observations and gave

$$c = 299,776 \pm 6 \text{ km/sec.}$$

The agreement of this value with that obtained by the most accurate work of rotating mirror is excellent.

Houston's Method

In 1949 R.A. Houston (Great Britain) reported the results of another optical method in which Piezo-electric effect was used. According to this effect if a quartz crystal is suitably cut and placed in H.F. alternating electric field it may be alternately compressed or extended thus generating H.F. oscillations. The method is again essentially modification of Fizeau's toothed wheel method where the wheel has been replaced by a Piezo-quartz crystal for alternate interruptions of the light beam of the order of 200 times as rapid as Fizeau's wheel.

The arrangement of the apparatus. Rays from a monochromatic source of light S after reflection from a glass plate M emerge through a convex lens L_1 as a parallel beam and then passing through a quartz plate Q fall on a screen X with a slit at A. If the quartz plate is subjected to an alternating electric field, high frequency oscillations will be generated in it and it will be crossed by at number of nodal planes perpendicular to the plane of paper, where the refractive index will be different from that at other places. In other words, the quartz plate will act like an intermittent diffraction grating with a frequency double the frequency of oscillations. Diffraction spectra will be produced. If the slit A in the screen X lies in the direction of the first order spectrum, the light passes through and is brought to focus by a converging lens L_2 on a concave mirror C. This light retraces its path and on arriving at Q if it finds the quartz plate exactly in the same condition as on its forward journey, may emerge in the direction QM and enter the eye which, therefore, sees an image of S.

Let f be the frequency of the oscillations then the rate of formations of the grating is $2f$/sec. Therefore when the image is seen by the eye it is evident that the time taken by the light to traverse twice the distance QC is some multiple of $1/2f$.

If the collimating system L_2C is now moved towards or away from the screen X in the direction QA, the image will soon disappear. It will, however, reappear when the shift d of L_2C is such that light takes $1/2f$ seconds in traversing $2d$. By taking several such positions, the mean value of d is determined, whence

velocity of light, $$c = \frac{2d}{1/2f} = 4df \qquad ...(5)$$

The value of the velocity of light in vacuum obtained by Houston in 1938 was found to be 299,725 km/sec. which agrees with values obtained from other accurate experiments.

Indirect methods. Several indirect methods depending upon the principles of electro-magnetism have been employed for measuring the velocity of light for which the interested reader should consult some books on current electricity. Mention, however, is made here of a few methods. In one of the methods used by Mercier in 1923 velocity of electric waves along wires is determined. High frequency electrical waves are sent along one of the two adjacent and parallel wires and back along the other to produce stationary waves. The positions of the nodes are located and by measuring the mean distance between two nodes and the frequency of oscillations, the velocity of light is calculated. Mercier's determination gave the velocity of light, c=299,782 ± 30 km/sec. in vacuum which is in excellent agreement with the results of the direct experiments.

Rosa and Dorsey in 1936 measured the ratio between electrostatic and electromagnetic units of electricity which according to Maxwell's electromagnetic theory is equal to the velocity of the electromagnetic waves. The result for c = 299,781 ± 10 km/sec. agrees remarkably well with other determinations.

In 1950 Essen making use of micro-waves of varying frequency and a cavity resonator gave the velocity of light in vacuum as 299,792.5 ± 30 km/sec.

Hansen and Bol working in United States of America used an improved cavity resonator method and found that the value of c in vacuo is equal to

$$299{,}789.9 \pm 0.4 \text{ km/sec.}$$

Value and Importance

The value and importance of the velocity of light. The present accepted value of the velocity of light in vacuum as a result of a large number of determinations by various experimenters is taken as

$$c = 299{,}790 \text{ km/sec.}$$

$$= 2.9979 \times 10^8 \text{ m/sec.}$$

It is a universal constant of great importance for the following reasons :

(i) *Constancy of velocity of light.* According to Einstein's relativistic optics this velocity is a universal constant. It is not attainable by any material bodies and is unaffected by the motion of the frame of reference. It is, therefore, being adopted as a standard of measurement in Geodetic surveys and for several other precision measurements such as frequencies of electromagnetic radiations and spectral lines by the use of the relation $v=c/\lambda$ where v is the frequency, c the velocity of light and λ the wavelength.

(ii) *Variation of mass with velocity.* The constancy of the velocity of light has led to the formulation of the present generally accepted theory of Relativity by Einstein. According to this theory the velocity of light enters into the expressions for mass, length and time values of different frames of reference moving relative to each other. For instance, the mass of a body moving with a velocity v is given by

$$m_v = \frac{m_o}{\sqrt{1 - \frac{v^2}{c^2}}}$$

where m_0 is she rest mass, c the velocity of light. According to this relation if $v = c$, mass is infinite.

Time and length magnitudes are also altered.

(iii) *Einstein's mass-energy relation.* According to this a mass m is equivalent to an amount of energy E and the equation- relating these quantities is $E = mc^2$, where c is the velocity of light. This relation has been verified during atomic fission and atomic fusion processes and thus enables us to calculate the amount of energy released during these processes.

(iv) *Energy momentum relation.* According to the Einstein mass energy relation radiant energy E is equivalent to a mass (E/c^2) moving with the velocity of light c and having a momentum

$$p = \frac{E}{c^2} \times c = \frac{E}{c} = \frac{hv}{c},$$

since $E = hv$, where h (= 6.625×10^{-24} joule-sec.) is Planck's constant and v the frequency of radiant energy. Hence a knowledge of c enables us to calculate easily the mass and momentum of a photon of radiation.

(v) *Relation between electromagnetic and electrostatic units.* The velocity of light enters into the relation between *e.m.* units and *e.s.* units which are connected by some multiple of it.

For instance,

e.m. unit charge = $c \times$ *e.s.* unit of charge.

e.m. unit of pot. diff.=(1/c) *e.s.* unit of pot. diff.

e.m. unit of capacity=$c^2 \times$ *e.s.* units of capacity,

and so on.

In fact these relations formed the basis of Maxwell's electromagnetic theory of radiation.

(vi) It has the same value in vacuum for all ethereal radiations such as wireless waves, X-rays, γ-rays, etc.

Recently, however, there has been some controversy about the possible variation in the velocity of light with time which is suggested by a study of the results obtained during the period from Cornu's experiments in 1874 to Pease and Pearson's in 1932. But such apprehensions have been set at rest for the time being by a careful analysis of all the significant experiments by Birge who has suggested for the velocity of light a value equal to

$$299{,}774 \pm 4 \text{ km/sec.}$$

2

Light Waves

Newton's chief objection to the wave theory lay in the fact that light travels in straight lines and does not bend round corners like sound waves. In a way the rectilinear propagation of light could find an easy explanation on the basis of Newton's corpuscular theory but was an apparent impossibility on wave theory. This was a stumbling block for the acceptance of the wave theory of light for a long time, as the simple exposition of the wave theory offered no explanation for it. It is a matter of common observation that waves of all kinds, we can think of, in the two dimensional demonstrations bend round the edge of an obstacle. In fact light as we shall see later in this chapter also bends round corners and that comparison with sound is a matter of degree. The small bending in case of light is on account of its very short wavelength which is of the order of 6×10^{-6} cm. The wavelength of sound is about 10^7 times as large as the wavelength of light so that while a small obstacle will cut off a beam of light we need comparatively a very large obstacle (10^7 times larger) to achieve the same effect in sound. This effect in the case of light was first of all discovered by

Grimaldi (Italy, 1618-1663). He observed that with a small source of light the shadow of an object was larger than that given by geometrical construction. He found that shadow was not sharp and well-defined but on close examination it was found to be surrounded by a system of (Iris coloured) fringes. These bands outside the geometrical shadow gradually get closer together until they fade away into uniform illumination. This phenomenon of bending of light waves around corners and their spreading into the geometrical shadow of an object is called 'diffraction' a name we owe to its discoverer.

Newton also observed this phenomenon but the explanation given by him was based on his corpuscular theory. Thomas Young later tried to explain the phenomenon as an interference effect between the direct light near the edge of the obstacle and the light reflected from the edge at grazing incidence. But Fresnel observed that this phenomenon did not depend upon the nature or shape of the edge of the obstacle. A round edge had the same effect as the sharp one, so he did not agree with Thomas Young that the phenomenon depended upon reflection. He conclusively proved by a series of experiments that this phenomenon was due to *mutual interference between the secondary wavelets from the same wavefront* more specifically, *it is the interference between the light coming from the various parts of a single aperture.* Huygens had already imagined these secondary wavelets as secondary sources whose wave-trains were *in a condition to interfere.* Since here not two but a very large number of sources (secondary wavelets) send out waves which all differ in phase when they reach the point under consideration, so mathematical treatment of the phenomenon is quite difficult.

Theory of Diffraction

The theory of diffraction is based upon the development by Fresnel of Huygens' view according to which every print in any surface through which light is passing may be considered as a secondary radiator and the subsequent effect beyond that surface may be ascribed to the combined action of the secondary wavelets alone. If this idea is correct then light from a sufficiently small aperture should spread in all directions beyond it. This is truly borne out by experiments.

Classification of diffraction phenomena. The diffraction phenomena which owe their appearance to a limitation of the wavefront can be classified into two main groups, named after the respective pioneers of these branches of the subject.

(a) *Fresnel diffraction.* It comprises shadow phenomenon in which interference takes place between light waves reaching a given point in or near a shadow from different parts of the same wavefront generating from an aperture without modification by lenses and mirrors. In this case either the source or the screen or both are at *finite* distances from the diffracting aperture or the obstacle. The wavefronts are either spherical or cylindrical. This class was investigated earlier.

(b) *Fraunhofer diffraction.* It includes all cases in which the source of light and screen or telescope in which the pattern is observed, are *effectively* at *infinite* distances from the intervening aperture or obstacle causing the diffraction or the diffracted light is focused by a lens or mirror giving the reduced image of the pattern that would appear on an infinitely distant screen in the absence of the lens or mirror. In this case interference occurs between sets of parallel rays coming past the aperture or obstruction in given directions. Fraunhofer class is simple to treat and easy to observe in practice. The light from a slit source is rendered parallel by a lens, so that the incident wavefronts are plane at the aperture and the secondary wavelets are initially all in phase. The diffracted light is focused on a screen with another lens placed on the other side of the aperture or obstacle. This lens forms in its focal plane a reduced image of the pattern that would appear on an infinitely distant screen in the absence of the lens.

The difference between the two classes may be further illustrated by comparing the forms of the wavefronts. From a source *S* falls on the aperture which allows a portion *AB* of the incident wavefront to pass through it. All the points in the plane of the diffracting aperture are not in the same phase of disturbance as the wavefront does not arrive at them simultaneously. *P* is a point on the screen where the effect is sought, so with this point as centre describe an arc *CD*. The effect at *P*, due to Fresnel diffraction is the resultant sum of all the disturbances which occur along CD at the same time. It is obvious that no actual wavefront exists at *CD* but it is possible to define the surface thus represented as the diffracted wavefront with respect to point *P*. In Fraunhofer diffraction which is depicted both the incident and the diffracted wavefronts *AB* and CD do actually exist and are plane, *AB* being rendered as such by the convex lens intervening the source and the aperture. All the points in the plane of the aperture are in the same phase of disturbance and give rise to a plane diffracted wavefront *CD* which is brought to focus on the screen at *P* with the help of another lens.

Broadly speaking, we can say that there is no essential difference in the nature of the diffraction process in the two cases and Fresnel diffraction merges gradually into Fraunhofer diffraction as the screen is moved away from the slit.

Fresnel's Class of Diffraction Phenomena

In the following sections, we shall confine ourselves to Fresnel's explanation of rectilinear propagation of light and a few cases of diffraction at obstacles and apertures by elementary methods.

***Fresnel's treatment of the wavefront* (1815).** We have explained earlier Huygen's principle of wave propagation. He attempted to explain the rectilinear propagation of light by means of wave envelopes alone. Thus if AB bean aperture, on which light is incident from a source S, he assumed that the effect of each element of the wavefront APB in producing the next wavefront $A_1P_1B_1$ was confined to the apex of the secondary wave or that part of it which touches the envelope. This practically amounted to the assumption of the law of rectilinear propagation of light itself.

The physical justification for the same was provided later by Fresnel who took into account the secondary wavelets originating from the various points of a wavefront and showed that the so-called rectilinear propagation of light was only approximate and the absence of light in the region of geometrical shadow was due to their mutual interference and thus neutralizing each other. To explain the phenomenon he made the following assumptions :

Fresnel's assumptions

1. A wavefront could be divided into a large number of elements or zones of small area and that each of these elements was a source of secondary waves which were effective everywhere.
2. The effect at any point O will be due to the resultant of the secondary wavelets reaching O from the various elements of the wavefront AB, the disturbance from P reaching O first and from the adjoining element on the wavefront a little later.
3. The amplitude of the disturbance at any point X due to the secondary waves from an element of the wave surface Situated at P is a function both of the distance PX and of the angle θ between the line PX and normal PO. It is maximum in a direction radially outward from P and falls to zero in the opposite direction. This variation with angle is called obliquity function or

inclination factor. In fact its effect along any direction is proportional to $(1 + \cos\theta)$. It is maximum along *PO*, since $\theta = 0$ and hence $\cos\theta = 1$; half along PA_1, cos 90° being zero and zero along *PS*, cos 180° being –1. Since the effect at the rear of the wave is zero, there will evidently be no *'back wave'*.

Rectilinear propagation of light-half period elements. Let us suppose that a point source of monochromatic light is sufficiently far away so that for all practical purposes, we are concerned with *plane wavefronts* only. Consider such a plane wavefront *ABCD* advancing towards some point *O*.

According to Huygens' principle of wave propagation each point on the wavefront *ABCD* becomes a centre of secondary disturbance and we have to find the resultant amplitude at the point *0* due to all these disturbances. All the points on the wavefront are in the same phase, but the secondary waves originating from it will reach *O* with different phases and different amplitudes. It is because the waves reaching *O* will have travelled different distances and the inclinations of their paths with normals at the respective points will also be different.

To determine the resultant effect at *O*, let the wavefront *ABCD* be subdivided into a number of elemental zones.

To do this, drop a perpendicular *OP* from *O* on the wavefront then *P* is known as the pole of the wavefront with respect to *O*. Let $OP = b$, λ be the wavelength of the monochromatic light constituting the wavefront. With *O* as centre and radii

$$OM_1 = \left(b + \frac{\lambda}{2}\right),$$

$$OM_2 = \left(b + \frac{2\lambda}{2}\right),$$

$$OM_3 = \left(b + \frac{3\lambda}{2}\right), \text{ etc.,}$$

describe a series of concentric spheres cut by the wavefront *ABCD* in circles of radii PM_1, PM_2, PM_3, etc. The areas enclosed between these circles etc., are known as Fresnel's half-period zones *or* half-period elements. The area enclosed by the first circle of radius PM_1 is called the first half-period zone. The areas of the annular strips M_1M_2, M_2M_3, etc., are known as second, third, etc., half-period zones.

Now the problem of finding the resultant effect of the whole wavefront at *O* reduces to finding out the resultant of a large number of disturbances originating from the annular strips into which the whole wavefront is divided. The *amplitude* due to wavelets produced by each zone reaching the point *O* depends on the following factors :

(*i*) It is *directly* proportional to the *area* of the zone. *The greater the area, the greater the number of ether particles sending out disturbances to the point O.*

(*ii*) It falls off *inversely* as the *distance* of the zone from the point *O*. *The greater the distance of the zone, the smaller is the amplitude of vibration reaching O.*

(*iii*) It varies with the *obliquity* of each zone, *i.e.*, with the angle between the normal to the wavefront and the line joining the point *O* to the zone. The *obliquity factor* is measured by the expression $(1 + \cos\theta)$. *The amplitude decreases with increasing obliquity.* This decrease is extremely slow but steady.

The resultant amplitude at *O* due to all the zones will also depend upon the *relative phases* of the wavelets from various zones producing the effect. These phases are related to the distances of the various zones from the point *O*.

To calculate the resultant amplitude at a point due to a plane wavefront let us consider each factor separately.

Area factor. It will now be proved that, to a first approximation the *areas of all zones are equal.*

The radius of the first half-period zone

$$PM_1 = \sqrt{OM_1^2 - OP^2}$$

$$= \sqrt{\left(b + \frac{\lambda}{2}\right)^2 - b^2}$$

$$= \sqrt{\left(b^2 + \frac{\lambda^2}{4} + b\lambda - b^2\right)}$$

$$= \sqrt{b\lambda} \text{ approx.}$$

neglecting $\frac{\lambda^2}{4}$ as compared to $b\lambda$.

Similarly, the radius of the second half period zone

$$PM_2 = \sqrt{OM_2^2 - OP^2}$$

$$= \sqrt{\left[(b+\lambda)^2 - b^2\right]}$$

$$= \sqrt{b^2 + \lambda^2 + 2b\lambda - b^2}$$

$$= \sqrt{2b\lambda} \text{ approx.}$$

More generally, the radius of $(n - 1)$th zone

$$PM_{n-1} = \sqrt{\left(b + \overline{n-1}\,\frac{\lambda}{2}\right)^2 - b^2}$$

$$= \sqrt{(n-1)\,b\lambda}$$

and the radius of the nth zone

$$PM_n = \sqrt{\left(b + n\frac{\lambda}{2}\right)^2 - b^2}$$

$$= \sqrt{nb\lambda} \text{ approx.} \qquad \text{... (1)}$$

The areas of the different zones may be calculated as under:

The area of the first half-period zone = $\pi PM_1{}^2 = \pi b\lambda$.

The area of the first two half-period zones = $\pi PM_2{}^2 = 2\pi b\lambda$.

$\therefore$ the area of the second half-period zone

$$= 2\pi b\lambda - \pi b\lambda = \pi b\lambda.$$

The area of the nth zone = area of the first n zones – area of the first $\overline{n-1}$ zones

$$= \pi PM_n^2 - \pi PM_{n-1}^2$$

$$= \pi\left(nb\lambda - \overline{n-1}\,b\lambda\right) = \pi b\lambda.$$

Thus, we find from equations (1) and (2) that *the radii of these half-period zones are proportional to $\sqrt{n}$ where $n = 1, 2, 3$, etc., while the areas of the zones are approximately equal* and independent of the order of the zone, so long as n is small. Because of this approximate equality of the areas of the different zones, the amplitudes of the secondary wavelets contributed by each zone are very nearly equal.

Distance and obliquity factors. With the increase in the order of the contribution made by each successive zone will continue to decrease regularly but slowly due to *distance* and *obliquity* factors *i.e.*, if m_1, m_2, m_3, etc., represent the amplitudes at O due to first, second, third, etc., half-period zones respectively, then m_1 is slightly greater than m_2, m_2 than m_3 .... m_{n-1} than m_n, such that *the amplitude of any zone may be taken as the arithmetic mean between the preceding and the succeeding zones.* Thus to a close approximation,

$$m_2 = \frac{m_1 + m_3}{2}$$

$$m_4 = \frac{m_3 + m_5}{2}$$

and so on. The difference between any two successive m's also becomes less and with the increase in their order because of their increasing narrowness as we go out from the first zone.

Relative phases of the zones. Since the distances of the half-period zones from the point O are different, the secondary waves originating from these zones will reach O in different phases.

(a) The phase of the wavelets coming from P is zero. The wavelets from M_1 will arrive $T/2$ seconds later than those from P as OM_1 is greater than OP by $\lambda/2$. The difference in phase, therefore, for wavelets coming from P and $M_1 = \pi$ radians. The phases of the wavelets from intermediate points between P and M_1 will vary from 0 to π. Hence the average phase of all the wavelets from the first half-period zone

$$= \frac{0 + \pi}{2} = \frac{\pi}{2} \text{ radians.}$$

(*b*) The phase difference between wavelets from M_1 and M_2 will again be π as they differ in path by $\lambda/2$ and hence that between wavelets from P and $M_2 = 2\pi$ radians, that is, the phase of the wavelets from the second half-period zone will vary between π and 2π. So the average phase of wavelets from second half period zone

$$= \frac{\pi + 2\pi}{2} = \frac{3\pi}{2} \text{ radians,}$$

which is *opposite* to that due to the wavelets from the first half period zone. Similarly, we can calculate that the phases of the wavelets from the third half-period zone vary between 2π and 3π so that the average phase of the wavelets from this zone is $5\pi/2$. Proceeding in this manner we find that the resultant phase difference between the wavelets from any two successive zones is π radians. Thus successive zones are in opposite phase so that the resultant amplitude at O due to the whole wavefront is given by

$$A = m_1 - m_2 + m_3 - m_4 + m_6 - \dots \pm m_n$$

according as n is odd or even

$$\text{or} \quad A = \frac{m_1}{2} + \left(\frac{m_1}{2} - m_2 + \frac{m_2}{2}\right) + \left(\frac{m_2}{2} - m_4 + \frac{m_5}{2}\right) + \frac{m_5}{2} \dots .$$

The expressions within brackets are each equal to zero and the resultant amplitude at O thus reduces to

$$A = \frac{m_1}{2} + \frac{m_n}{2}, \text{ if } n \text{ is odd}$$

$$\text{and} \qquad A = \frac{m_1}{2} + \frac{m_{n-1}}{2} - m_n, \text{ if } n \text{ is even.} \quad \dots (3)$$

If n is quite large then the effect due to the $(n - 1)$th or nth is almost negligible on account of the distance and obliquity factors and the *resultant amplitude due to the whole wavefront reduces to $m_1/2$, that is, one-half of that which will be produced by the first half period zone alone.*

As the intensity I is proportional to the square of the amplitude so $I = m_1^2/4$, *i.e.*, intensity at O due to the whole wavefront is one-fourth of intensity due to the first half-period element.

If the number of exposed zones is plotted along the X-axis and the intensity along the Y-axis, then the variation of intensity at O is graphically represented by the curve.

From this graph it is obvious that only a few central zones are effective and the intensity at O due to them is practically the same as due to the entire wave front. If the obstacle is large enough to obstruct the effective central zones, the entire wavefront would become ineffective.

As shown above, since λ (= 6×10^{-5} cm) is extremely small the area $\pi b\lambda$ of each half-period zone is very small*, so that even a small obstacle placed at P would cut off a considerable number of central effective half-period elements and the resultant intensity at O, being one-fourth of the intensity due to the first exposed zone will be almost negligible, *i.e.*, no light will be received at O. Thus a small obstacle at P will screen O almost entirely from the wavefront and this is what we mean when we say that light is propagated in straight lines. In fact approximate rectilinear propagation of light is a consequence of extreme shortness of the wavelength of light, aided with Fresnel's ingenious way of dividing a wavefront and introducing the idea of destructive interference between the light waves from neighbouring half-period elements.

Since the wavelength of sound waves for the ordinary frequencies of human voice (20 to 20,000) is almost a million times larger than the wavelength of light waves so that the area of the first half-period zone ($\pi b\lambda$) for the sound waves will be very large. Thus to stop the effect of a sound wavefront, an obstacle of about a million times bigger in size is required. Hence for an obstacle of ordinary size sound waves do not appear to travel in a straight line which is consistent with the wave theory.

Zone Plate

An interesting example of diffraction and also a verification of the foregoing theory is found in the zone plate. The effects due to the consecutive zones in the wavefront *ABCD* (Art. 14-4), being in antiphase destroy each other so it is possible to obtain very bright illumination at O by covering up alternate zones. If b be the distance of the point O from the pole P of the wavefront and λ the wavelength of the monochromatic light employed, then the radii of the various zones are given by

$$r_1 \sqrt{1\, b\lambda}\ ,\ r_2 = \sqrt{2\, b\lambda}\ ,\ r_3 = \sqrt{3\, b\lambda} \text{ and } r_n = \sqrt{nb\lambda}$$

Thus the radii of the zones are proportional to the square roots of the natural numbers, *i.e.*, $\sqrt{1}, \sqrt{2}, \sqrt{3},$ etc. It is this fact on which the construction of the zone plate depends. To construct a zone plate, therefore, draw on a sheet of white card-board piece a set of large number of concentric circles with radii proportional to the square roots of the natural numbers. Paint the alternate zones, say even, black. Take a photograph of this pattern on a reduced scale. This forms the *zone plate.* It will give a system of areas corresponding to the half-period elements, having odd zones transparent and even zones opaque when the zone plate is said to be positive. If the even zones are transparent, the zone plate is called negative.

Theory of the zone plate. Let us consider AA an imaginary transparent plate perpendicular to the plane of the paper, and S, a point source of monochromatic light giving out spherical waves of wavelength A. Draw SP perpendicular to the plane AB and produce it to O. Let us suppose that we arc required to find out the resultant displacement at the point O on the screen placed at right angles to the plane of the paper.

With P as centre imagine the plate AB to be divided into circular zones with radii equal to PM_1, PM_2, PM_3 etc., such that

$$SM_1 + M_1O = SP + PO + \lambda/2$$

$$SM_2 + M_2O = SP + PO + 2\lambda/2$$

....

$$SM_n + M_nO = SP + PO + n\lambda/2$$

To determine the radius r_n, say, of the nth circle, consider the relation

$$SM_n + M_nO = SP + PO + n\lambda/2. \qquad \text{... } (i)$$

Putting $SP = a$, $PO = b$, we have

$$SM_n = \sqrt{SP^2 + PM_n^2} = \sqrt{a^2 + r_n^2}$$

$$= a\left(1 + \frac{r_n^2}{a^2}\right)^{\frac{1}{2}} = a + \frac{r_n^2}{2a} \text{ approx.}$$

by binomial theorem since r_n is very small as compared to a, and the terms containing higher powers of $\frac{r_n^2}{a^2}$ are neglected.

Similarly, $$M_nO = \sqrt{PO^2 + PM_n^2} = \sqrt{b^2 + r_n^2}$$

$$= b + \frac{r_n^2}{2b} \text{ approx.}$$

Substituting the values in relation (*i*) above, we have

$$a + \frac{r_n^2}{2a} + b + \frac{r_n^2}{2b} = a + b + \frac{n\lambda}{2},$$

or $$r_n^2\left(\frac{1}{a} + \frac{1}{b}\right) = n\lambda, \quad \text{... } (ii)$$

whence $$r_n{}^3 = n\lambda\left(\frac{ab}{a+b}\right) \quad \text{... } (4)$$

Since a, b, and λ are constants.

$$\therefore \quad r_n \propto \sqrt{n}.$$

It thus follows that *the external radii of the various zones for given values of a and b are proportional to the square root of the natural numbers.*

The area of the nth zone

$$= \pi\left(r_n^2 - r_{n-1}^2\right)$$

$$= \pi\left[\frac{n\lambda ab}{a+b}\right] - \frac{(n-1)\lambda ab}{a+b}$$

$$= \pi\frac{\lambda ab}{a+b} \quad \text{... } (5)$$

which is independent of the order n. This means that for the given positions of point source and its image the *areas of all the zones remain practically constant.* But as m_1, m_2, m_3 , etc., the displacements at O due to the various zones, decrease very slightly with the order of the zones and are alternately in *opposite* phase, the resultant amplitude is given by

$$A = m_1 - m_2 + m_3 - m_4 + \ldots..$$
$$= m_1/2$$

when n is very large.

If monochromatic light from S is allowed to fall upon this zone plate and the emergent light is received on a screen, then for a certain value of b the hypothetical half-period elements of the incident wavefront may coincide with the actual elemental areas of the plate. In such a case if the even number of elements are made opaque, then the resultant displacement

$$A_1 = m_1 + m_3 + m_5 + \ldots \quad \ldots (6)$$

which is many times greater than that due to all zones. This explains the *focusing* action of a zone plate. Under these conditions O can be said to be the image of S.

Putting equation (*ii*) in the form $\frac{1}{a}+\frac{1}{b} = \frac{n\lambda}{r_n^2}$. ... (7)

and comparing it with the convex lens formula

$$\frac{1}{v}+\frac{1}{u} = \frac{1}{f}, \text{ we have}$$

$$\frac{1}{f} = \frac{n\lambda}{r_n^2}, \quad \therefore \quad f = \frac{r_n^2}{n\lambda} \quad \ldots (8)$$

which gives the primary or first order focal length of the zone plate.

Multiple foci of a zone plate. A zone plate has a number of foci of diminishing intensity as we go from the brightest focus above mentioned towards the plate. They are determined by the condition that for each of these foci an actual zone contains an *odd* number of half-period elements of the incident plane wavefront. This is due to the reason that the area of each half-period element diminishes as b decreases. The positions of the foci are evaluated as follows : The *primary* first order focal length, as it is called, is equal to the value of b in equation (7) above when $a = \infty$;

$$\text{whence} \qquad f = \frac{r_n^2}{n\lambda}$$

$$\text{or} \qquad f_1 = \frac{r_1^2}{\lambda} \qquad \text{... (8a)}$$

For this focal length each actual zone of the plate contains one half-period element and the focal point O_1, say, is called the first order focal point, the intensity of the image is maximum being proportional to $(A_1)^2$. *cf.* equation (6).

If each zone contains three half period elements with respect to some other nearer point O_2, the alternate three half-period elements being obstructed then again the intensity will be maximum giving rise to a second focus of intensity less than that in the first case, since the resultant amplitude at O_2 will be equal to

$$A_2 = (m_1 - m_2 + m_3) + (m_7 + m_8 + m_9) + (m_{12} - m_{14} + m_{15}) + \ldots\ldots$$

$$= \left(\frac{m_1}{2} + \frac{m_3}{2}\right) + \left(\frac{m_7}{2} + \frac{m_9}{2}\right) + \left(\frac{m_{13}}{2} + \frac{m_{15}}{2}\right) + \ldots..$$

$$= \frac{1}{2}(m_1 + m_3 + m_7 + m_9 + \ldots..) \text{ approx.}$$

$$< A_1.$$

The light rays transmitted through each successive transparent zone have a mean path difference of 3λ. Hence O_2 is sometimes called the third order focal point.

As in this case the first transparent zone of the plate includes three half-period elements all of equal area, hence if x be the radius of the first new half-period element, we have

$$3\pi x^2 = \pi r_1^2$$

or $$x^2 = \frac{r_1^2}{3}.$$

whence we have for the second focal length

$$f_2 = \frac{x^2}{\lambda} = \frac{r_1^2}{3\lambda} = \frac{f_1}{3}.$$

The third focus occurs at a position, when each zone of the plate contains five half-period elements, the alternate 5 half-period elements being blocked off. The resultant displacement is given by

$$A_3 = (m_1 - m_2 + m_3 - m_4 + m_5) + (m_{11} - m_{12} + m_{13} - m_{14} + m_{15}) + \ldots$$
$$= \left(\frac{m_1}{2} + \frac{m_5}{2}\right) + \left(\frac{m_{11}}{2} + \frac{m_{15}}{2}\right) + \ldots.$$
$$= \frac{1}{2}\left(m_1 + m_5 + m_{11} + m_{15} + \ldots..\right)$$
$$< A_2.$$

Since in this case the first transparent zone includes five half-period elements, we have

$$5\pi y^2 = \pi r_1^{\,2}$$

or $$y^2 = \frac{r_1}{5},$$

where y is the radius of the first new half-period element,

whence $$f_3 = \frac{y^2}{\lambda} = \frac{r_1^2}{5\lambda} = \frac{f_1}{5}$$

and so on.

A zone plate has thus multiple foci of focal length $f_1, f_1/3, f_1/5$, intensities in decreasing order. Generalizing the above result, the focal lengths associated with various foci can be obtained from the equation,

$$p = \frac{r_1^2}{(2p+1)\lambda}$$

where $$p = 0, 1, 2, 3, \text{etc.}$$

A zone plate and a convex lens compared

(*i*) Both produce real image of an object on the side other than the object.

(*ii*) Both show chromatic aberration as the focal length varies with l for both.

(*iii*) Points of difference between a convex lens and zone plate :

(a) For a given wavelength a convex lens has *one* focus, given by the relation

$$\frac{1}{f} = (\mu - 1)\left(\frac{1}{r_1} - \frac{1}{r_2}\right)$$

where letters have their usual significance, while a zone plate has a *number* of foci of diminishing intensity as shown above and situated at $f_1, f_1/3, f_1/5$ etc., where $f_1 = \frac{r_1^2}{\lambda}$.

(*b*) The intensity of the image in a zone plate decreases as the order of the focal length increases as shown above.

(*c*) In a zone plate the red focus is nearer *P* than violet, *i.e.*, $f_r < f_v$ since $f = \frac{r_n^2}{n\lambda}$ and $\lambda_r > \lambda_v$. In case of a lens it is just the reverse, red rays being less refrangible than violet come to focus farther away from the lens, *i.e.*, $f_r > f_v$.

Example. *For a wavelength of light $\lambda = 6 \times 10^{-5}$ cm and radius of the first half-period zone r = 0.6 mm, a zone plate brings rays to focus at its brightest spot. Find the focal length of the equivalent lens.*

$$f = \frac{r_n^2}{n\lambda} = \frac{r_1^2}{\lambda}$$

$$= \frac{(0.06 \text{ cm})^2}{6 \times 10^{-5} \text{ cm}} = 60 \text{ cm.}$$

Intensity at a point due to cylindrical wavefronts. Consider a long and narrow slit *SS*, held perpendicular to the plane of the paper illuminated by a monochromatic source of light of wavelength λ. The envelope of the secondary wavelets spreading out from the slit is cylindrical with the slit as the axis of the cylinder.

Consider a horizontal section *XPY* of the cylindrical wavefront emanating from the slit with its axis passing through the mid-point of the slit and lying in the plane of the paper-the slit *S* will in turn pass through the surface *XPY*.

To determine the effect of this wavefront at *O*, on the screen situated ahead of this wavefront, let it be divided into *elementary half-period strips* with respect to the point *O*, in a manner similar to Fresnel's half-period zones of a plane or spherical wavefront. Join the mid-points of the slits to *O* cutting the wavefront at *P*, then *P* is the pole of the wavefront with respect to the point *O* on the screen. With *O* as centre, and $(OP + \lambda/2)$ as radius describe an arc cutting the wavefront in M_1 and N_1, PM_1 and PN_1 represent the sections of narrow strips on the cylindrical surface *XPY*. PM_1 is termed the *first help-period element* of the surface *PX* and PN_1 that of the surface *PY*. Similarly, with *O* as centre and radii $(OP + 2\lambda/2)$, $(OP + 3\lambda/2)$, etc., draw half-period strips for each half, *PX* and *PY*, of the wave surface cutting them in M_2, M_3, etc., and N_2, N_3, etc., respectively. This gives us the strips of width PM_1, M_1M_2, M_2M_3 .... and PN_1, N_1N_2, N_2N_3 .... each parallel to the slit. These are respectively called first, second, third etc., half-period strips. Their appearance when viewed from the point *O*. This method of dividing is called *strip division* of the wavefront. The width of each succeeding strip is less than the one preceding it decreasing rapidly at first and more slowly latter on. Since the length of these strips remains constant, the areas of the half-period strips are proportional to their widths and are not equal but in decreasing order. The effects contributed by them at *O* are also in decreasing order and alternately in the same phase.

If m_1, m_2, m_3 be the numerical values of the displacements at O due to the wavelets from first, second, third, half-period elements on either side of the pole the resultant effect R at O due to either of the wave is given by

$$R = m_1 - m_2 - m_3 - m_4 + m_5 - .... \pm m_n$$

$$= \frac{m_1}{2} + \left(\frac{m_1}{2} - m_2 + \frac{m_3}{2}\right) + \left(\frac{m_3}{2} - m_4 + \frac{m_5}{2}\right) + ... \pm \frac{m_n}{2}$$

$$= m_1/2,$$

since the expressions within the brackets reduce to zero and $m_n/2$ is negligible when n is very large.

Thus the resultant effect at O due to the entire wave surface is $2R = m_1$.

Fresnel diffraction of a cylindrical wavefront at a straight edge. Let A, be a sharp and straight edge of an opaque obstacle AB (edge of a razor blade) and S a narrow rectangular slit both parallel to each other and perpendicular to the plane of the paper. Let the slit be illuminated by a monochromatic source of light of wavelength λ. Join SA and produce it to meet the screen CD placed perpendicular to the plane of the paper at O. Let us determine the distribution of light intensity on the screen.

According to the laws of geometrical optics the shadow of the edge should start at the point O in the downward direction, while the illumination should start abruptly at O in the upward direction, *i.e.*, below O there should be total darkness and above O total illumination. But in actual practice we observe that :

1. Below O the intensity falls off continuously and *rapidly* as we move into the geometrical shadow until complete darkness is reached a short distance below O.
2. Just above O alternate dark and bright bands of unequal

width and varying intensity parallel to the straight edge are observed. The bands become closer and less distinct as we move away from O and ultimately they disappear and we get uniform illumination.

We shall now proceed to explain these observations on the basis of the wave theory.

Intensity variation inside the geometrical shadow. Let XY be the trace of a cylindrical wavefront S when it reaches the diffracting straight edge at A. For the point O on the screen the pole of the wavefront is at A and only the upper thalf AX of the wavefront it exposed so that the displacement at O is only one-half of what it would have been if whole of the wavefront were effective. Consider this half wavefront to be divided into half-period strips. Let the amplitudes of the waves sent by them to point O be m_1, m_2, m_3, etc. Remembering that successive strips are in opposite phase, the resultant amplitude of all the waves is

$$A = m_1 - m_2 + m_3 - m_4 + ...$$

$$= \frac{m_1}{2} \qquad ... (9)$$

Hence the intensity at O is proportional to $\left(\frac{m_1}{2}\right)^2 = \frac{m_1^2}{4}$.

As we move to a point below O, into the geometrical shadow, say to Q_1, the pole also moves from A to P_1, and gradually the first, the first two, the first three, etc., half-period elements of the upper half of the wave are also successively cut off, so that if only the first half period element is obstructed the displacement at Q_1 is given by

$$A = -m_2 + m_3 - m_4 + m_5 - .....$$

$$= -\frac{m_2}{2}.$$

Similarly, if two or three half-period elements are

obstructed, the displacements at O are $\frac{m_3}{2}, \frac{m_4}{2}$ respectively and so on. As the intensity is proportional to the square of the displacement, the magnitudes of the intensities beginning with O downwards are $\left(\frac{m_1}{2}\right)^2, \left(\frac{m_2}{2}\right)^2, \left(\frac{m_3}{2}\right)^2$ etc. As m_1, m_2, m_3, m_4, etc., are in the descending order of magnitude, therefore, as the point moves into the geometrical shadow the intensity of light falls off rapidly and soon diminishes, to zero.

Explanation of diffraction bands outside the geometrical shadow. As we move from O towards the point Q distant y from O, the pole of the wavefront also moves up towards P and one after another the first, and first two, the first three, etc., half-period strips or elements of the lower half of the wavefront are also exposed so that the illumination at the point Q will be due to the complete upper half of the wavefront above P together with that due to the half period elements contained in the part of the wavefront PA If the number of half-period elements in PA be even, the displacements from the alternate elements mutually cancel each other out in pairs and the intensity at Q is due only to the upper half of the wavefront above P. If the number of half-period elements contained in PA is odd, one element will be left over uncancelled and the effect of this at Q will be added to that of the half wavefront above P so that the intensity will be greater.

The displacement at O is due to upper half wave and is equal to $\frac{m_1}{2}$; that at a point a little above O so that one element of the lower half wave is exposed it is equal to $\frac{m_1}{2} + m_1 = \frac{3m_1}{2}$. Similarly, for a point further up when two elements of the lower half wave are exposed the displacement $= \frac{m_1}{2} + m_1 - m_2$ $= \frac{3m_1}{2} - m_2$ and so on. The corresponding intensities of light are $\left(\frac{m_1}{2}\right)^2, \left(\frac{3m_1}{2}\right)^2, \left(\frac{3m_2}{2} - m_2\right)^2$, etc. Thus as the point Q moves up from O it passes alternately through the positions of *maximum* and *minimum* brightness according as PA contains odd or even number of half-period elements. This gives rise to what are known as diffraction bands. The intensity of a maximum goes on decreasing and that of a minimum goes on increasing and finally Q gets into uniform illumination.

Mathematical Treatment

To find the positions of the diffraction bands and their widths let us again consider the effect at a point Q outside the geometrical shadow. Join SQ cutting the wavefront at P. Also join AQ. The number of half-period elements contained in PA is equal to the number of half wavelengths in the path difference $(BP - PQ)$.

Putting $SA = a$

and $AO = b$, we have

$$AQ = \sqrt{AQ^2 + OQ^2}$$

$$= \sqrt{b^2 + y^2} = b\left(1 + \frac{y^2}{b^2}\right)^{\frac{1}{2}}$$

$$= b + \frac{y^2}{2b} \text{ approx.}$$

...[by binomial theorem

and $PQ = SQ - SP$

$$= \sqrt{(a+b)^2 + y^2} - a$$

$$= (a+b)\left(1 + \frac{y^2}{(a+b)^2}\right)^{\frac{1}{2}} - a$$

$$= (a+b)\left(1+\frac{y^2}{2(a+b)^2}+\ldots\ldots\right)-a \ldots.$$

... [by binomial theorem]

$$= a+b+\frac{y^2}{2(a+b)}-a$$

$$= b+\frac{y^2}{2(a+b)}.$$

Hence the path difference

$$AQ - PQ = \left(b+\frac{y^2}{2b}\right)-\left(b+\frac{y^2}{2(a+b)}\right)$$

$$= \frac{y^2}{2}\left(\frac{1}{b}-\frac{1}{a+b}\right) = \frac{ay^2}{2b(a+b)}.$$

For maximum displacement at Q this path difference must be equal to an odd multiple of $\lambda/2$ and for minimum displacement it should be equal to an even multiple of $\lambda/2$.

$\therefore$ for maximum brightness

$$\frac{ay^2}{2b(a+b)} = (2n+1)\frac{\lambda}{2}$$

or
$$y = \sqrt{\frac{b(a+b)(2n+1)\lambda}{a}}, \quad \ldots (10)$$

and for minimum brightness

$$\frac{ay^2}{2b(a+b)} = 2n\frac{\lambda}{2},$$

or
$$y = \sqrt{\frac{b(a+b)2n\lambda}{a}} \quad \ldots (11)$$

where n is whole number, 0, 1, 2, 3 etc.

Since y is proportional to $\sqrt{2n+1}$ for a maximum and $\sqrt{2n}$ for a minimum, the width of bands goes on decreasing with the increase in the order of the number of the band, *i.e.*, as Q moves away from O.

Diffraction Pattern

The graph depicts the intensity distribution due to diffraction at the straight edge. In the geometrical shadow the intensity falls off rapidly till within a short distance of the shadow of the edge there is complete darkness. Outside the geometrical shadow there is a system of dark and bright bands till uniform illumination results.

If white light is used in place of monochromatic light, a few coloured bands running parallel to the shadow of the edge will be observed in the illuminated region above O. This is due to the reason that white light is composite in character and the width of the band due to each colour will be different. All these effects are observed in practice using a straight edge placed at about 2 meters from the slit, the screen being a similar distance away on the other side of the edge. Using monochromatic light and noting the positions of the first and the one most distant and clearly visible, say nth bright band by a travelling microscope we can find the wavelength of light using the relation (10) derived above.

Distinction between interference and diffraction. The following important points of distinction may be noted between the interference and diffraction phenomena at this stage :

(1) Interference phenomenon is due to the interaction of light waves resulting from the division of a wavefront or of amplitude into *two parts* primarily originating from the same source while diffraction is due to the interaction of light waves from different parts of the same wavefront.

(2) The interference bands are equally spaced ($\beta = D\lambda/d$), whereas the diffraction bands go on decreasing in width as we move away from the edge of the shadow $\left(y \propto \sqrt{2n}\right)$.

(3) The intensity distribution of light between a maximum and a minimum in interference bands varies between zero and maximum while in the case of diffraction bands maximum and minimum amplitudes are superposed on a background of uniform illumination due to the upper half of the wave so that the intensities of the minima are comparable to those of the maxima hence the bands have a poor contrast.

Example. *A narrow slit illuminated by light of wavelength 6×10^{-5} cm is placed at a distance of 10 cm from a straight edge. If measurements are done at a distance of 1 meter from the straight edge, calculate the distance between the first and the second dark bands.*

The distance of the nth dark band outside the geometrical shadow is given by

$$y_n = \frac{\sqrt{b\left(a+b\right)2n\lambda}}{a}$$

Here, $a = 10$ cm ; $b = 100$ cm ; $\lambda = 6 \times 10^{-5}$ cm.

For the first dark band $n = 1$, so that

$$y_1 = \sqrt{\frac{2 \times 100\left(10+100\right) \text{cm} \times 1 \times 6 \times 10^{-5} \text{ cm}}{10}}$$

For the second dark band $n = 2$, so that

$$y_2 = \sqrt{\frac{2 \times 100\left(10+100\right) \text{cm} \times 1 \times 6 \times 10^{-5} \text{ cm}}{10}}$$

$$= 0.5187 \text{ cm}$$

Hence the distance between the first and the second dark bands

$$y_2 - y_1 = 0.5137 \text{ cm} - 0.3634 \text{ cm} = 0.1503 \text{ cm}.$$

Fresnel diffraction at a narrow obstacle. Let AB narrow obstacle (generally a wire) of thickness d, held parallel to a narrow rectangular slit S placed perpendicular to the plane of the paper. If the slit is illuminated by monochromatic light of wavelength A, it will give rise to a *cylindrical wavefront* XY falling on the wire.

The presence of the wire in the path of light produces a geometrical shadow MN on the screen EF held normal to the plane of the diffraction bands outside M and N and also sometimes inside the geometrical shadow interference bands are observed depending upon the thickness of the wire.

(a) *Thin wire.* When the obstacle is a thin wire, and obstructs a few half-period elements of each half of the wave it behaves just like two sharp edges so that at any point R_1 or R_2 outside the geometrical shadow the effect is same as due to a straight edge at A or B. The diffraction bands of *unequal* intensity and *decreasing* width will be seen above M and below N on both sides of the limits of the geometrical shadow similar to those discussed in the preceding section. These fringes are called *exterior fringes* and are obviously due to diffraction. They are independent of the thickness of the wire because on either side, the effect, of the other part of the wavefront is negligible—important half-period zones being cut off. Thus the upper system is formed by the upper half of the wavefront and the lower system by the lower half.

Besides these exterior diffraction fringes there are also *interior* interference fringes formed inside the geometrical shadow by the superposition of the waves bending into the shadow from both sides. The effect of the upper portion AX of the wavefront may be regarded as that due to a small luminous source at A and that of lower portion BY of the wavefront equal to a small luminous source at B, *i.e.*, A and B act as two coherent sources of light.

The effect at any point Q within the geometrical shadow will depend upon the path difference $(BQ - AQ)$. Since the wavelets from A and B start in the same phase and will also meet at O in phase so that the centre O will always be bright. The point Q will be bright or dark according as $(BQ - AQ)$ is even or odd multiple of $\lambda/2$, *i.e.*,

for Q to be bright $BQ - AQ = n\lambda$, and

for Q to be dark $BQ - AQ = (2n + 1)\lambda/2$.

If d be the thickness of the wire AB, D the distance of point O from P the mid-point of the wire, $OQ = y$ then for the nth bright band at Q, we have by the theory of interference fringes,

$$y = \frac{D}{d} n\lambda$$

Also if β be the fringe-width of the interference fringes, then

$$\beta = \frac{D}{d} \lambda$$

These bands are equi-spaced and have almost uniform intensity of illumination.

It may be noticed that these interior fringes are ordinary Young's *interference fringes*. They can be distinguished from the exterior diffraction bands by covering one edge of the wire, in which case the interior interference fringes disappear but the diffraction bands on the other side remain.

We can calculate the diameter of the wire measuring β by a micrometer eye-piece and knowing D and λ.

(b) *Thick wire.* When the diameter d of the wire is increased, the fringe-width of the interference bands inside the geometrical shadow decreases and when the wire becomes sufficiently thick, the interior fringes vanish, for the waves cannot bend to such an extent as to superimpose and produce interference within the geometrical shadow. The intensity falls off gradually and rapidly as we enter the geometrical shadow from M or N side.

It may be clearly understood that the fringes with white light will be coloured. The patterns beyond M and N would remain the same as in a thin wire.

Example. *A very narrow vertical slit illuminated with sodium light of wavelength* 6.0×10^{-5} *cm casts a shadow of a vertical steel wire of diameter 2 mm and one meter away on a screen 3 meters away from the wire. Calculate the distance apart of bands inside the shadow and the total number of bands which can be seen.*

It may be noted that in the geometrical shadow DC equidistant bright and dark interference bands will be observed. The fringe-width of these interference bands is given by

$$\beta = \frac{D}{d}\lambda.$$

Here, $D = 300$ cm, $d = 0.2$ cm.

$$\lambda = 6.0 \times 10^{-5} \text{ cm. } 300 \text{ cm.}$$

$$\therefore \quad \beta = \frac{300 \text{ cm}}{0.2 \text{ cm}} \times 6.0 \times 10^{-5} \text{ cm.}$$

$$= 0.09 \text{ cm.}$$

To find the total number of bands we are required to find out the width of the geometrical shadow CD. From the diagram it is clear that

$$\frac{CD}{AB} = \frac{SO}{SP},$$

or $$CD = \frac{SO}{SP} \times AB$$

$$= \frac{(100 + 300)\text{ cm}}{100\text{ cm}} \times 0.2\text{ cm} = 0.8\text{ cm}.$$

Total number of bands

$$= \frac{0.8\text{ cm}}{0.09\text{ cm}} = 9\text{ approx.}$$

Thus there will be nine bands—central bright band surrounded by four bright bands on either side.

Fresnel diffraction at a rectangular aperture (slit). Let S be a narrow rectangular slit source of monochromatic light, placed perpendicular to the plane of the paper and AB a rectangular aperture held parallel to it, XY is the section of a cylindrical wavefront issuing from S and incident on the aperture.

According to the laws of geometrical optics the edges A and B of the aperture will cast the geometrical shadows above C and below D on the screen EF held normal to SO, the line joining the source and the mid-point of the aperture and there should be brightness in the region CD.

(a) *Wide aperture.* When the aperture is wide and contains many half-period elements, say, about one hundred, with respect to a point on the screen, the edges A and B, of the aperture behave like two separate edges independent of each other. Within the geometrical shadow on each side, *i.e.,* above C and below D, the intensity of illumination falls off gradually and rapidly but just inside the geometrical image near C and D there will be found diffraction bands parallel to the edges of the aperture exactly similar to those produced by a straight edge. These bands will be of unequal width and non-uniform intensity. As we move away from C and D towards O, the mid-point of CD, the bands go on decreasing in width and visibility. The general illumination increases till it becomes uniform near about O.

(b) *Narrow aperture.* When the aperture AB is narrow and contains only a few half-period elements with respect to the middle point O of the geometrical image CD on the screen, the illumination at O will be *maximum* or *minimum* according as the number of half-period elements contained in each half of the exposed wavefront above and below the pole P with respect to the point O is *odd* or *even*. For instance, for a particular position of the screen let us suppose that the number of zones is odd, say, three in each half of the wave, the displacement at O

$$= 2\,(m_1 - m_2 + m_2)$$

$$= 2\left[\left(\frac{m_1}{2} - m_2 + \frac{m_3}{2}\right) + \left(\frac{m_1}{2} + \frac{m_3}{2}\right)\right]$$

$$= 2\left(\frac{m_1}{2} + \frac{m_3}{2}\right)$$

$$= m_1 + m_3 \text{ and so a maximum.}$$

If the number of zones is *even*, say, four in each half of the wave then the displacement at O

$$= 2\,(m_1 - m_2 + m_3 - m_4)$$

$$= 2\left[\left(\frac{m_1}{2} - m_2 + \frac{m_3}{2}\right) + \left(\frac{m_1}{2} - m_4 + \frac{m_3}{2}\right)\right]$$

$$= 0, \text{ approximately and so a minimum.}$$

We thus see that the *illumination at O has a maximum or a minimum value according as the aperture comprises an odd or even number of half period elements in each half of the wave surface.*

Keeping the screen fixed in this position if we now move to a point Q within the geometrical image CD, the illumination is maximum or minimum according as the exposed part in each half of the wavefront above and below the new pole contains odd or even number of half-period elements with respect to Q. For example, if there are three half-period elements exposed in the upper half of the wavefront and five in the lower half, the resultant displacement at Q

$$= (m_1 - m_2 + m_3) + (m_1 - m_2 + m_3 - m_4 + m_5)$$

$$= \left(\frac{m_1}{2} + \frac{m_3}{2}\right) + \left(\frac{m_1}{2} + \frac{m_5}{2}\right) = m_1 + \frac{m_3}{2} + \frac{m_5}{2},$$

giving a maximum intensity at Q. If the point Q moves still higher, there may be two exposed half-period elements from the upper-half of the wavefront and six from the lower half, then the resultant displacement at Q

$$= (m_1 - m_2) + (m_1 - m_2 + m_3 - m_4 + m_5 - m_6)$$

$$= m_1 - \frac{m_4}{2} - \frac{m_6}{2}$$

$$= 0, \text{ approx.}$$

giving a minimum intensity at Q. Similarly, for a still higher position of Q there may be one exposed element from the upper half and seven from the lower half of the wavefront, the displacement at Q

$$= m_1 + (m_1 - m_2 + m_3 - m_4 + m_5 - m_6 + m_7)$$

$$= m_1 + \frac{m_1}{2} \left(\frac{m_1}{2} - m_2 + \frac{m_3 2}{}\right) + \left(\frac{m_3}{2} - m_4 + \frac{m_6}{2}\right)$$

$$+ \left(\frac{m_5}{2} - m_6 + \frac{m_7}{2}\right) + \frac{m_7}{2}$$

$$= m_1 + \frac{m_1}{2} + \frac{m_7}{2},$$

giving a maximum intensity.

For the point C the pole of the wavefront is at A and all

the tight elements of toe lower-half only are exposed so that the displacement at C

$$= m_1 - m_2 + m_3 - m_4 + m_5 - m_6 + m_7 - m_8 ...$$

$$= \frac{m_1}{2} + \left(\frac{m_1}{2} - m_2 + \frac{m_3 2}{2}\right) + \left(\frac{m_3}{2} - m_4 + \frac{m_5}{2}\right)$$

$$+ \left(\frac{m_5}{2} - m_6 + \frac{m_7}{2}\right) + \frac{m_7}{2} - m_8 ...$$

$$= \frac{m_1}{2} + \frac{m_7}{2} - m_8,$$

giving a minimum intensity.

Thus the upper part *OC* of the geometrical image *CD* is crossed by dark and bright bands. Similarly, it can be shown that the lower part *OD* is also crossed by dark and bright bands. Thus a set of alternately dark and bright bands parallel to the edges of the geometrical shadow are seen inside the geometrical image *CD*. These are evidently diffraction bands being due to the mutual interference of the disturbance from half-period elements of the same wavefront.

Now let us see what happens when the screen is gradually moved away from the slit. We have seen above that when the number of zones in each half of the wave is four, then the displacement at *O* is *minimum*. When the screen is gradually moved away from the aperture, the number of zones in *AB* with respect to it gradually decreases and the illumination at *O* will undergo a change in intensity. If the number in each half of the wave decreases to three, the displacement at *O* is maximum (*cf.* page 339). If the screen is moved still further away from the slit the number of zones in each half will become two and the displacement at *O* can be shown to be minimum in the same way and so on, till the slit comprises one or less than one half-period element in each half of the wave surface with respect to the point *O* on the screen, when there will be a broad flat maximum at the centre as discussed in case (*c*) below.

Let us now find the effect at any point R in the geometrical shadow, when the pole of the wavefront shifts to a point above A. In this case upper half of the wavefront is totally cut off and the point R receives energy only from some of the zones of the lower half of the wavefront and the intensity of illumination at R will be maximum or minimum according as AB contains odd or even number of half-period elements with respect to R. If the path difference

$$BR - AR = (2n + 1)\ \lambda/2,$$

i.e., an odd multiple of $\lambda/2$ there will be an odd number of half-period elements in AB and so R will be a *bright* point, due to an excess of one half-period element after the mutual cancellation of other half-period elements in pairs.

If the path difference

$$BR - AR = 2n\ \frac{\lambda}{2},$$

i.e., an even multiple of $\lambda/2$ there will be an even number of half-period elements in AB and this will make the point R *dark* due to mutual cancellation of half-period element in pairs.

To discuss it a little more in detail if the path difference $(BR - AR) = \lambda$, so that the path difference between the waves from A and P is $\lambda/2$ and also between those from P and B is $\lambda/2$, for every point in the upper half of the slit there is a corresponding point in the lower half for which the path difference is $\lambda/2$. Such waves on reaching R destroy each other making the net displacement at R due to the whole slit AB as zero.

With the path difference $(BR - AR) = 3\lambda/2$, the slit can be divided into three equal parts so that the wavelets from the two adjoining parts on reacting R cancel out each other and those from the third part produce brightness at R.

When the path difference $(BR - AR) = 2\lambda$, the slit can be considered to be divided into four equal parts so that the wavelets from the alternate parts on reaching R mutually cancel out each other and the net displacement at R is zero.

Thus the bands within the geometrical shadows beyond C and D are also *diffraction bonds* of decreasing intensity and width. The bands will be narrower than those within the region CD. With the increase in distance from the edges of the shadow, there is uniform darkness. Thus there is no *marked transition* between the types of bands inside the geometrical image and outside it unlike the case of diffraction at a narrow obstacle in which case the bands inside the geometrical shadow are interference bands and outside the geometrical shadow diffraction bands.

If $OR = y$, $AB = d$ and $PO = D$, we shall have

$$y = \frac{D}{d}(2n + 1)\frac{\lambda}{2} \quad \text{for a } \textit{bright} \text{ band}$$

and

$$y = \frac{D}{d} n\lambda \text{ for a dark band.}$$

The distance between two bright bands is equal to $D\lambda/d$ so that they are equi-spaced and a decrease in the width of the aperture increases the width of a band.

(c) *A very narrow aperture.* When the aperture comprises one, or lens than one, half-period element in each half of the wave surface with respect to the point O on the screen there will be a broad flat maximum at the centre, the illumination remaining considerable far out into the geometrical shadow on each side. There will be no bands inside the geometrical image CD. But for any other point R far into the region of the geometrical shadow there is a possibility of interference due to the disturbances from different parts of the slit giving rise to well-defined bands formed in the manner already described. Fresnel diffraction at a small aperture, circular aperture. Let S, be a point source of monochromatic light of wavelength λ, AB a *small* circular aperture such as a pin hole and O a point on the screen on the axial line SPO. Let XY be the trace of a spherical wavefront at any instant touching the circular aperture.

Suppose we want to study the intensity at O.

To do this divide the aperture AB into *Fresnel's half-period zones* with respect to O in the same manner. The intensity at O will depend upon the size of the aperture and its distance PO from the point under consideration.

If the aperture comprises an even number of half-period zones with respect to O, they mutually cancel out each other in pairs, resulting in *almost* zero intensity at O. But if the aperture contains an odd number of half-period zones, then after mutual cancellation with each other in pairs one zone is left over and there will be some intensity at O. Hence as O travels along the axis of the aperture, the intensity of illumination at it passes. The distance of any point of maximum or minimum intensity from the aperture can be easily calculated.

If $SP = a$, $PO = b$, and $PA = PB = r$, the radius of the aperture, then the path difference between the marginal and central rays is given by

$$\begin{aligned}
\delta &= (SA + AO) - (SP + PO) \\
&= (a^2 + r^2)^{1/2} + (b^2 + r^2)^{1/2} - (a + b) \\
&= a\left(1 + \frac{r^2}{a^2}\right)^{1/2} + b\left(1 + \frac{r^2}{b^2}\right)^{1/2} - (a + b) \\
&= \left(a + \frac{1}{2}\cdot\frac{r^2}{a} + b + \frac{1}{2}\frac{r^2}{b}\right) - (a + b) \\
&= \frac{r^2}{2}\left[\frac{1}{a} + \frac{1}{b}\right] = \frac{r^2}{2} \times \frac{a + b}{ab}.
\end{aligned}$$

$$\therefore \qquad r^2 = \frac{2ab\delta}{a+b}$$

$$\pi r^2 = \frac{2\pi ab\delta}{a+b}.$$

If this area contains n half-period zones

$$\delta = n\lambda\ /2,$$

$$\therefore \qquad \pi r^2 = \pi abn\lambda/(a + b)$$

whence $$b = ar^2/\ (na\lambda - r^2), \qquad ...(13)$$

so that when b changes, n alternately changes from odd to even and the intensity at O successively passes through maximum and minimum values, the dark points corresponding to even values of n and the bright points to odd values.

If the incident light is parallel, the area of each half-period zone is $\pi b\lambda$ (Art. 14-4) and the number of half-period zones in the aperture is equal to $\pi r^2/\pi b\lambda$; when b changes, this number changes. If this number is odd, we get a maximum at O, if this is even, a minimum.

Let us discuss a specific case when the screen is kept fixed and the aperture is small so that it exposes only four half-period zones with respect to O. This displacement at O will be

$$= m_1 - m_2 + m_3 - m_4 = \frac{m_1 + m_3}{2} - m_4,$$

which is a minimum value since m_1 m_3 and m_4 are nearly equal. The point O will, therefore, be dark.

For a point on the screen between O and E, the pole of the wave surface will be eccentric with respect to the aperture. If the distance of this point from O is small, about half the fourth zone becomes intercepted on the upper side and half the fifth zone becomes exposed on the lower side of the aperture. The displacement is, therefore

$$= m_1 - m_2 + m_3 - \frac{m_4}{2} + \frac{m_5}{2}$$

$$= \left(\frac{m_1 + m_3}{2}\right) - \frac{m_4}{2} + \frac{m_4}{2}$$

$$= \frac{(m_1 + m_3)}{2} \text{ nearly,}$$

which is a maximum value and corresponds to a bright fringe. Since the intensity will be uniform at all points at the same distance from O the bright fringes will be circular.

At a point still further from O half the third zone will become intercepted at the upper side and half of the sixth zone exposed at the lower side of the aperture. The displacement will then be

$$= m_1 - m_2 + \frac{m_3}{2} - \frac{m_4}{2} + \frac{m_5}{2} - \frac{m_6}{2}$$

$$= \frac{m_1}{2} - \frac{m_6}{2} \text{ nearly,}$$

which is a minimum value and corresponds to a dark circular fringe and so on.

Thus the point O is surrounded by a series of circular bands of unequal widths, alternately bright and dark. In case of a large aperture these bands are visible only near the limits of geometrical shadow as in the case of a straight edge with uniform illumination in the centre. Inside the geometrical shadow, however, the intensity falls off continuously and rapidly until complete darkness is reached a short distance from the boundary of the shadow.

A very narrow aperture

If the aperture is so small that it contains only a fraction of the first half-period zone with respect to O, the point O will be *bright* and there will be no bands within the geometrical image of the aperture on the screen. With respect to a point on the screen beyond the geometrical image, the aperture exposes a number of short lengths which are parts of succeeding half-period zones. This point will, therefore, be bright or dark according as the aperture contains an odd or even number of these fractional zones with respect to it. Thus when the aperture is very small, its geometrical image will be surrounded by a large number of alternately dark and bright circular bands lying within the geometrical shadow. The distance between the centres of the two neighbouring bright bands is given by

$$\frac{b}{2r}\lambda, \qquad \text{...(14)}$$

Example. *A circular aperture of 1.2 mm diameter is illuminated by plane waves of monochromatic light. The diffracted light is received on a distant screen which is gradually moved towards the aperture. If the centre of the circular patch of light first becomes dark when the screen is 30 cm from the aperture find the wavelength of light used.*

For the first darkness on the axis, the aperture should contain first two half-period zones only. Since the area of each half-period zone for plane waves is $\pi b\lambda$ and the area of the circular aperture is πr^2. For first darkness, we have

$$\frac{\pi r^2}{\pi b\lambda} = 2, \quad \text{or} \quad r^2 = 2b\lambda, \quad \text{or} \quad \lambda = \frac{r^2}{2b}.$$

Here, $r = 0.12 \text{ cm}/2 = 0.06 \text{ cm}, \; b = 30 \text{ cm}$

$$\therefore \qquad \lambda = \frac{(0.06 \text{ cm})^2}{2 \times 30 \text{ cm}} = 6 \times 10^{-5} \text{ cm}.$$

Fresnel diffraction at an opaque circular disc. Let S, be the point source of monochromatic light of wavelength λ, AB the opaque circular disc and O a point on the screen on the axial line SPO, where P is the pole of the wavefront XY at any instant.

To consider the effect at O divide the wavefront XY into a number of Fresnel's half-period zones with respect to O. Depending upon its size, the disc will obstruct some of the central half-period zones. The resultant displacement at O will be equal to half of the displacement due to the first of the series of the un-obstructed zones.

If the size of the disc is so small that it obstructs only the first half-period zone with respect to O, the displacement at O will be

$$= \quad - m_2 + m_3 + m_4 + ... = - \frac{m_2}{2}.$$

Hence the intensity of the light is proportional to $(- m_2/2)^2$.

If the disc covers up first two zones, the resultant displacement at O

$$= \quad + m_3 — m_4 + m_5 - ... \frac{m_3}{2}$$

and the intensity is proportional to $(m_3/2)^2$, which is not much different from that in the first case. If the disc is not to be there, the resultant displacement would have been $= m_1 - m_2 + m_3 - ...... = m_1/2$ and the intensity proportional to $m_1{}^2/4$, which is only slightly greater than $m_2{}^2/4$ or $m_3{}^2/4$. Hence we find that when the disc obstructs only a few central zones, O is always bright. This remarkable result was first of all deduced by Poisson, but was confirmed experimentally by Arago who proved that the result of the experiment was in complete agreement with the predictions of wave theory of light. When the point O is quite close to the disc the obliquity factor very much reduces the intensity of illumination at O.

If the size of the disc increases so that it cuts off a large number of effective half-period zones the intensity of illumination is reduced accordingly, but *the point O will always remain bright provided the size of the disc is very small.*

At any other point in the geometrical shadow CD of the disc the diffracted waves meet with a phase difference and interference pattern results. Thus the point *0* will be surrounded *by fine* circular interference bands *within* the geometrical shadow *CD* of the disc. These bands are similar in origin to those found within the shadow of a narrow rectangular obstacle such as a wire. These are interference bands. But just outside there will bo *broad* circular diffraction bands bordering the geometrical shadow *CD*. They are of *unequal* width, similar to those produced by a straight edge.

The authors performed the experiment by allowing sunlight to enter a darkened room through a pin hole in the door and let it fall on an Indian 'pie' coin suspended by a thread at a distance of about 5.5 meters from the pin hole. A shadow of the pie was obtained on a screen at a similar distance on the other side ; a clearly *well-defined* bright spot was seen in the centre of the shadow surrounded by faint circular bands due to interference. This experiment confirms Fresnel's approach to the whole problem of diffraction of light.

The Integrals

Fresnel's integrals. For evaluating the intensity of illumination at any point of a diffraction pattern, Fresnel developed some special forms of integrals which we shall now consider. Let us take the case of a cylindrical wavefront *XY*, emanating from a long narrow slit *S* perpendicular to the plane of the paper. Let it be divided into a system of narrow strips by lines drawn on the surface of the wave- front parallel to the axis of the cylinder. It is convenient to replace each of these strips by its central effective portion so that the whole is reduced to an equatorial band coinciding with *XY* in the plane of the paper normal to the slit. *P* is the pole of the wave surface with respect to a point *O* where the effect is sought. Let *a* be the radius of the circle *XY* and *b* the distance of *O* from *P*. The secondary wavelets from the various elements like δ s of the wavefront *XY* travel different distances to *O*, that is, they arrive at *O* with different phases. To find the resultant amplitude at O due to the wavelets from XY, let ωt be the phase of the disturbances arriving simultaneously at all points on XY. Then the phase of the wavelet from P which is at O at time t will be $(\omega t — \beta b)$ where $b = PO$ and $\beta = 2\pi/\lambda$ and the phase of the disturbance at O from an element at M at the same time will be $(\omega t - \beta t)$ where $c = MO$.

To obtain an expression for the disturbance δD at O due to an elementary strip of width δD situated at M, we may suppose that it is directly proportional to the product of δs and $\cos\{\omega t - \beta(b + z)\}$ where $z = c - b$. Neglecting the obliquity factor and the decrease of amplitude with distance, we may write approximately,

$$\delta D = r \cos\{\omega t - \omega b) - \beta z)\}\ \delta s,$$

where r is a constant.

Or we may write,

$$\delta D = \{r \cos(\omega t - \omega b) \cos \omega z\}\ \omega s + \{r \sin(\omega t - \beta b) \sin \beta z\}\ \delta s$$

Summing up the effect of all the elements between P and M we may write in the language of integral calculus

$$D = r \cos(\omega t - \beta b)$$

$$\int_0^s \cos \beta z\, ds + r\sin(\omega t - \beta b) \int_0^s \sin \beta z\, ds$$

$$= R \cos(\omega t - \beta b - \phi), \qquad ...(i)$$

where $R \cos \phi = r \int_0^s \cos \beta z\, ds + R \sin \phi$

$$= r \int_0^s \sin \beta z\, ds \qquad ...(ii)$$

Thus we conclude that the resultant disturbance has the same period as the wave but a different amplitude R and a different initial phase.

The resultant intensity is proportional to R^2, which is given by

$$R^2 = \left[\left\{\int_0^s \cos \beta z\, ds\right\}^2 + \left\{\int_0^s \sin \beta z\, ds\right\}\right] \quad ...(iii)$$

To express z in terms of s, consider the triangle SOM. By the cosine law, we have

$$c^2 = (a + b)^2 + a^2 - 2a\,(a + b) \cos \theta$$

$$= (a + b)^2 + 2a\,(a + b) \left[1 - \frac{\theta^2}{2} + ...\right]$$

$$... \text{ if } \theta \to 0$$

$$= b^2 + a\,(a + b)\,\theta^2 = b^2 + \frac{a+b}{a}\, s^2$$

$$... \left[\because \theta = \frac{s}{a}\right.$$

$$= b^2 \left[1 + \frac{a+b}{a} \cdot \frac{s^2}{b^2}\right]$$

$$\therefore \quad c = b\left[1 + \frac{a+b}{a} \cdot \frac{s^2}{b^2}\right]^{1/2}$$

$$= b\left[1 + \frac{a+b}{a} \cdot \frac{s^2}{b^2}\right], \text{ ...if } \frac{a+b}{a} \cdot \frac{s^2}{b^2} \to 0$$

$$= b + \frac{a+b}{2ab} \cdot s^2$$

$$\therefore \quad z = c - b = \frac{a+b}{2ab} \cdot s^2 \quad ...(iv)$$

Putting $\beta z = \frac{1}{2} \pi v^2$, where v is a new variable,

we have $\frac{2\pi}{\lambda} \cdot \frac{1+b}{2ab}\, s^2 = \frac{1}{2}\pi v^2, ... \because \beta = \frac{2\pi}{\lambda}$ ...(v)

$$\text{or } s = v\sqrt{\frac{ab\lambda}{2(a+b)}}, \text{ so that } \delta s = \delta v\sqrt{\frac{ab\lambda}{2(a+b)}}$$

and substituting these values in *(iii)*, we get

$$R^2 = \frac{ab\lambda}{2(a+b)}\left[\left\{\int_0^v \cos\left(\frac{1}{2}\pi v^2\right) dv\right\} + \left\{\int_0^v \sin\left(\frac{1}{2}\pi v^2\right) dv\right\}^2\right]$$

$$= \text{constant} \times (P^2 + Q^2) \quad \text{...(vi)}$$

Taking the constant as unity, the intensity

$$I = R^2 = P^2 + Q^2.$$

The integrals, $P = \int_0^v \cos\left(\frac{1}{2}\pi v^2\right) dv$ ...(15)

and $Q = \int_0^v \sin\left(\frac{1}{2}\pi v^2\right) dv$ ...(16)

are known as Fresnel's integrals.

To evaluate these integrals we proceed by the method of integration by parts, and get,

$$\int_0^v \cos\frac{1}{2}\pi v^2\, dv = M\cos\frac{1}{2}\pi v^2 + N\sin\frac{1}{2}\pi v^2 \quad \text{...(17)}$$

$$\text{and } \int_0^v \sin\frac{1}{2}\pi v^2\, dv = M\sin\frac{1}{2}\pi v^2 - N\cos\frac{1}{2}\pi v^2 \quad \text{... (18)}$$

where $M = \frac{v}{1} - \frac{\pi^2 v^5}{1.3.5} + \frac{\pi^4 v^4}{1.3.5.7.9} \dots \dots$ ...(19)

and $N = \frac{\pi v^2}{1.3} - \frac{\pi^3 v^7}{1.3.5.7} + \frac{\pi^5 v^{11}}{1.3.5.7.9.11} \dots \dots$ ...(20)

These series are divergent and are adapted for calculations when v is small. Note further that each integral after passing through a series of maxima and minima rapidly attains stationary value as shown in table on the next page, after Gilbert.

Cornu's Spiral

In 1874 A. Cornu interpreted Fresnel's integrals in an elegant graphical method for investigating intensity distribution in Fresnel's diffraction patterns. When the values of the Fresnel's integrals expressing the intensity are known then if we put

$$x = \int_0^v \cos\left(\frac{1}{2}\pi v^2\right) dv \text{ and } y$$

$$= \int_0^v \sin\left(\frac{1}{2}\pi v^2\right) dv$$

Table of Fresnel's integrals (Gilbert)

v	$\int_0^v \cos\frac{1}{2}\pi v^2\,dv$	$\int_0^v \sin\frac{1}{2}\pi v^2\,dv$	v	$\int_0^v \cos\frac{1}{2}\pi v^2\,dv$	$\int_0^v \sin\frac{1}{2}\pi v^2\,dv$
0.0	0.0000	0.0000	2.6	03389	0. 5500
0.1	0.0999	0.0005	2.7	0.3926	0. 4529
0.2	0.1999	0.0042	2.8	0.4675	0.3915
0.3	0.2994	0.0141	2.9	0.5624	0.4102
0.4	0.3975	0.0334	3.0	0.6057	0.4963
0.5	0.4923	0.0647	3.1	0.5616	0.5818
0.6	0.5811	0.1105	3.2	0.4663	0.5933
0.7	0.6597	0.1721	3.3	0.4057	0.5193
0.8	0.7230	0.2493	3.4	0.4385	0.4297
0.9	0.7648	0.3398	3.5	0.5326	0.4153

Contd...

v	$\int_0^v \cos\frac{1}{2}\pi v^2\,dv$	$\int_0^v \sin\frac{1}{2}\pi v^2\,dv$	v	$\int_0^v \cos\frac{1}{2}\pi v^2\,dv$	$\int_0^v \sin\frac{1}{2}\pi v^2\,dv$
1.0	0.7799	0.4383	3.6	0.5880	0.4923
1.1	0.7638	0.5365	3.7	0.5419	0.5750
1.2	0.7154	0.6234	3.8	0.4481	0.5656
1.3	0.6386	0.6863	3.9	0.4223	0.4752
1.4	0.5431	0.7135	4.0	0.4984	0.4205
1.5	0.4453	0.6975	4.1	0.5737	0.4758
1.6	0.3655	0.6383	4.2	0.5417	0.5632
1.7	0.3238	0.5492	4.3	0 4494	0.5540
1.8	0.3363	0.4509	4.4	0.4383	0.4623
1.9	0.3945	0.3734	4.5	0.5258	0.4342
2.0	0.4883	0.3434	4.6	0.5672	0.5162
2.1	0.5814	0.3743	4.7	0.4914	0.5669
2.2	0.6362	0.4556	4.8	0.4338	0.4968
2.3	0.6268	0.5525	4.9	0.5002	0.4351
2.4	0.5550	0.6192	5.0	0.5636	0.4992
2.5	0.4574	0.6197	∞	0.5000	0.5000

and plot them against each other in the Cartesian coordinates using the values given in the above table we obtain the curve, known as *Cornu's spiral.* It denotes the vibration curve for a cylindrical wavefront. The radius vector r drawn from the origin to any point on the curve will represent the resultant vibration arising from the corresponding portion of the wave. For we have $r^2 = x^2 + y^2$ and consequently r^2 represents the intensity of illumination and r represents the amplitude of the resultant vibration, and the angle which it makes with the axis of X, represents its phase.

On examining the properties of the curve we get some useful results as given below:

1. The distance measured along the spiral from the origin O to a given point (x, y) is equal to v, for if we differentiate the integrals x and y, we have

$$\delta x = \cos \frac{1}{2}\pi v^2 \; \delta v$$

and

$$\delta y = \sin \frac{1}{2}\pi v^2 \; \delta v.$$

Now the distance between two neighbouring points (x, y) and $(\delta + \delta x, y + \delta y)$ is given by

$$\sqrt{(\delta x)^2 + (\delta y)^2} = \delta v.$$

Hence the length of the arc of the curve measured from 0 to (x, y) is

$$\int dv = v \qquad \text{...(21)}$$

which proves the result.

2. The slope of the curve at the point (x, y) is

$$\frac{dy}{dx} = \frac{dy}{dv} \div \frac{dx}{dv} = \frac{\sin\frac{1}{2}\pi v^2}{\cos\frac{1}{2}\pi v^2}$$

$$= \tan \frac{1}{2}\pi v^2 . \qquad \text{...(22)}$$

Hence if x and y are drawn on the same scale (Cornu's spiral), the angle which the tangent to the curve at the point (x, y) makes with OX is

$$\frac{1}{2}\pi v^2 = \frac{2\pi}{\lambda} \cdot \frac{a+b}{2ab} \; S^2$$

$$= \frac{2\pi}{\lambda} \; z = \phi, \text{ say,}$$

which is the phase difference corresponding to a path difference z.

Hence $$v = \sqrt{\frac{2\phi}{\pi}} \qquad ...(23)$$

3. You will note that the values of v which are indicated on the curve by 1, $\sqrt{2}$, $\sqrt{3}$ correspond to points for which the phase differences are equal to $\pi/2$. π, $3\pi/2$, etc., respectively for at each of these points the tangent is parallel to one or other of the axes OX and OY.

4. The coordinates of the end points, J_1, and J_2, of the spiral are $\left(\frac{1}{2}, \frac{1}{2}\right)$ and $\left(+\frac{1}{2}, -\frac{1}{2}\right)$ respectively. The distance OJ_1 is $1/\sqrt{2}$ and it denotes the amplitude contributed by the upper half of the wave and, similarly, for OJ_2. The intensity due to the whole wave is

$$(OJ_1 + OJ_2)^2 = \left(\frac{1}{\sqrt{2}} + \frac{1}{\sqrt{2}}\right)^2 = 2.$$

This is a proportionality number, when the constant in equation *(vi)*, $I = R^2 = \text{constant} \times (P^2 + Q^2)$ is unity. Hence if we wish to express the intensity I as a function of the unobstructed intensity I_0, we get

$$\frac{I}{I_0} = \frac{R^2}{2} \qquad ...(vi)$$

Application of Cornu's spiral to simple cases of diffraction. From the preceding discussion it is clear that each half of a cylindrical wavefront gives rise to a spiral curve and these spirals are identical in shape and position. Thus OJ_1 represents the whole effect of the upper half and OJ_2 that of the lower half so that J_1J_2 gives the effect of the whole of the wave front. Let OM be the effect of a portion of the lower half, then the effect of this combined with the effect of the whole of the upper half gives J_1M as the resultant.

We shall now explain the intensity variation in a few cases already discussed with the help of Cornu's spiral.

Straight edge. We have seen that when light passes a straight edge of an opaque obstacle, a system of fringes is observed just outside the geometrical shadow but that inside the shadow the intensity fades away gradually. The illumination outside the geometrical shadow, say at Q is due to the complete half wave PX and a portion PA of the lower-half wave.

The effect of the upper half-wave is represented by OJ_1, an that of the portion PA of the lower half by OM, where M is some point on the lower half of the spiral. The resultant effect is vectorially represented by J_1M. Hence as M, moves along the lower spiral, J_1M passes through a maximum and a minimum for every convolution of M round it. This accounts for the bright and dark fringes outside the geometrical shadow. The values of maxima are greater than J_1J_2 which is the amplitude of the disturbance due to the unobstructed wave and those of minima being greater than OJ_1 that due to half wave but less than J_1J_2. At the edge of the geometrical shadow the amplitude is OJ_1 and the intensity is consequently one-fourth of that produced by the whole wave, since $I \propto a^2$.

For every point within the geometrical shadow one-half of the spiral is cut off completely and the illumination is due to a part of the spiral OJ, as only a fraction of half a wave is now in action. As a point on the screen recedes within the shadow, the extremity of the amplitude factor recedes from O along the spiral OJ moving round towards the point J. The amplitude JM, where M is a tracing point on the convolutions of the spiral decreases steadily and finally becomes zero without passing through any alterations.

Narrow aperture. The light energy reaching the point O of the screen is due to the elemeñt δx of the wavefront in the width *AB*.

Now, it was shown that

$$\delta s = \delta v \sqrt{\frac{ab\lambda}{2(a+b)}}$$

whence $$\delta v = \delta s \sqrt{\frac{2(a+b)}{ab\lambda}}$$

an equation which represents the length of an arc of the spiral corresponding to the elements $\Delta\ s$ of the wavefront. The resultant amplitude at the mid-point of the image of the aperture is then given by the chord of the spiral, the arc of which is $\Delta\ v$ and it will extend from *G* to *H*, on the spiral. In general, inside the geometrical projection of the aperture, the arc in question passes through the point *O* and belongs partly to one-half of the spiral and partly to the other. Outside the projection, the arc is situated altogether in one-half of the spiral when the length *R* of the corresponding chord gives the amplitude required.

When the slit is very wide we get two straight edges and the discussion in case (*i*) above applies to each edge. In the centre of the image there are very minute variations and close to the edge of the geometrical shadow marked changes in intensity occur.

When the slit is narrow so as to contain a few half-period elements in its each half, fringes appear both within and outside the geometrical image. $\Delta\ v$ is of such a length that a segment comprises numerous convolutions in each branch of the spiral.

When the slit is very narrow, the corresponding arc of the spiral is also small and it must move a considerable distance along either branch of the spiral before it curls so much that the chord is shortened. In this case there is a broad flat maximum at the centre which extends far beyond the confines of the geometrical image on either side.

Example. *At what distance from a slip of width 2c is it necessary to place a screen so that the central fringe may be of minimum intensity? What is this intensity ? (Carnu).*

We have to determine points on the two branches of the spiral which are nearest for the intensity of the central fringe to be minimum. Since in this case they are symmetrically situated, the chord joining them passes through O and with a compass the minimum length of the chord is found to correspond to $v = \pm\ 1.875$. Since s is equal to c of our problem, we have

$$\frac{2(a+b)}{ab\lambda}\ c^2 = (1.875)^2.$$

Again the intensity is measured by the square of the chord of the curve which measures 1.08, so the intensity = $(1.08)^2$ = 1.17. Now the intensity due to the whole wave is $J_1J_2^{\ 2} = 2$, the *relative* intensity of the central band is, therefore, 1.17/2= **0.585.**

Narrow wire. The case of a narrow wire may be regarded as complement of a rectangular aperture of similar dimensions. Inside the geometrical shadow the portions of each half of the wavefront being equivalent in effect to, say, arcs OM_1 and OM_2 on the two spirals have been stopped and the effects of the remaining portions are represented by J_1M_1 and J_2M_2. If the wire obstructs a considerable number of half-period elements from each half of the wave, the arcs OM_1 and OM_2 will contain several convolutions of the spiral and the lines J_1M_1 and J_2M_2 will be nearly equal so that there will be a *minimum* when J_1M_1 and J_2M_2 are in *opposite* directions and a *maximum* when in the *same* direction. In the region of the shadow, therefore, we have a system of *interference bands.* For points close to the borders of the shadow a portion, say, M_1M_2 is removed from the spiral but it is no longer placed symmetrically with respect to O. If

N_1N_2 be such a segment, the vectors J_1 N_1 and J_2 N_2 differ considerably in magnitude, so that when they are parallel and in opposite directions there will still be some resultant illumination and minima will not be places of complete darkness:

Outside the shadow we have a complete half wave and a few elements from one side of the wire. Their effects are equal to OJ_1 and OM_2 respectively, while from the other side we have a portion whose effect is represented by J_2M_2' where M_2' is a point on the spiral close to J_2. The are M_2M_2' is absent and of a constant length, proportional to the width of the wire. If J_2K, be the chord corresponding to this arc, then since J_1J_2 represents the effect of the unobstructed wave and J_2K indicates the effect of the intercepted portion, it is obvious that J_1K will represent the transmitted portion and will thus represent the resultant effect at a point outside the shadow and as the point recedes from the edge of the shadow, *i.e.*, as J_2K revolves about J_2, the line J_1K will pass through a series of maxima and minima which represent the external fringes. Now that there is marked transition between the internal and external fringes whereas the former are interference bands and the latter diffraction bands.

In the case of the narrow aperture, it is J_1K that is intercepted and J_2K transmitted.

3

The Diffraction

Fraunhofer Class of Diffraction Phenomena

It has already been pointed out that the diffraction effects in which the source of light and the screen on which the pattern is observed, are *effectively* at infinite distances from the aperture or obstacle concerned, belong to the Fraunhofer class of diffraction phenomena. These phenomena are easy to observe and simple to treat because here we have to deal with incident wavefronts which are plane at the aperture, so that the secondary wavelets are initially in the same phase. The incident light from a source is rendered parallel by a collimating lens and after passing through the diffracting aperture it is again focused on the screen by another lens This type of arrangement effectively removes both the source and the screen to infinity. In our discussion of this class of diffraction phenomena we shall consider a few problems which can be investigated by the graphical method.

Fraunhofer and Fresnel diffraction compared. The graphical methods applied to Fresnel class of diffraction phenomena where the wavefront is either spherical or cylindrical, may be used in describing Fraunhofer class as well. But in the latter case, since the diffracted wavefront is plane, the analytical methods for summing up the disturbances at a point are rather easier, and hence generally employed.

The experimental study of the *interference effects* of two or more beams of light bring out clearly the advantages of the Fraunhofer over Fresnel diffraction. These particular interference effects are without exception produced by devices belonging to class A in which a *division of* wavefront is made, consequently diffraction effects are superposed over them interfering with visibility. In all apparatuses of the Fraunhofer class such as the grating, the echelon and the Michelson's stellar interferometer *relatively* narrow sections of the wavefront from a slit source or a source of small size are utilized. They are focused with the help of a lens for observing the interference effects more conspicuously. This point may be illustrated by Fresnel diffraction through two slits. It is obvious that the effects due to the superposition of two wavefronts occurs in the middle—a region of poor illumination. There is no doubt that by decreasing the width of each slit the images of the slits could be made to spread out until they overlapped but it would result in no gain as illumination would decrease.

If a lens system is used as in Fraunhofer class the most intense portion of the diffraction patterns may be superposed at the focus, and the interference fringes will be observed in the brightest part of the field. This is the advantage in Rayleigh's refractometer over Young's experimental arrangement. The two cases are analogous, but the former uses Fraunhofer and the latter Fresnel diffraction. In Young's apparatus, the two slits producing the interfering beams are so close together that it cannot be used for comparison of optical path as needed in measuring the refractive indices of gases, etc.

Fraunhofer diffraction at a single slit. Let a parallel beam of light be incident normally upon a narrow slit AB of width a, where it gets diffracted. If a converging lens L is placed in the path of the diffracted beam, a real image of the diffraction pattern is formed on the screen in the focal plane of the lens. All the rays start from the slit AB in the same phase, they reinforce each other and produce brightness at a point O on the axis of the slit as they arrive there in phase. The intensity of the diffracted beam will be different in different directions and there will be some directions in which there will be no light. In fact it would depend upon the difference in distance from the two edges of the slit to the point under consideration. There will be a central bright band flanked symmetrically on *both sides* by a series of alternate dark and bright fringes whereas from the consideration of the geometrical optics of lenses we should expect a perfect image of the slit at O.

Let us consider the intensity at any point P_1 above O on the screen where another set of rays diffracted at an angle θ have been brought to focus by the lens L but the contributions from different elements of the slit AB do not arrive in phase. From A drop AE perpendicular to the diffracted ray from B, then AE constitutes the diffracted wavefront and BE is the path difference between the rays from the two edges A and B of the slit. Let this path difference be one wavelength. If now we imagine the aperture AB to be divided into two equal halves, then the wavelets from *corresponding points* in each half will differ in phase at P_1 by $\lambda/2$. They would mutually interfere and cancel out each other's effect thus producing the *first* minimum at P_1. For this to occur, therefore,

$$BE = \lambda.$$

Since $$BE = AB \sin\theta = a \sin\theta.$$

$\therefore$ $$a \sin\theta = \lambda$$

or $$\sin\theta = \frac{\lambda}{a}$$

or $$\theta = \frac{\lambda}{a} \quad ...(1)$$

since the angle of diffraction θ is usually very small.

If for another point P_2 the path difference BE is 2λ, then imagining the slit to be divided into four equal parts the rays from corresponding points separated by a distance $a/4$ in the two halves of each half of the slit will have a path difference of $\lambda/2$ and will mutually interfere and cancel out each other. A *second minimum,* therefore, occurs at P_2 in a direction θ given by

$$BE = a \sin\theta = 2\lambda$$

In general, the various minima will occur when the path difference\BE between the extreme rays is an even multiple of $\lambda/2$ or an integral multiple of λ, *i.e.,*

$$BE = a \sin\theta = 2n\frac{\lambda}{a} = n\lambda. \quad ...(a)$$

The minima on the other side of O are symmetrically situated at P', P'', etc.

Besides the central maximum at O, there are *secondary maxima* which lie in between the minima on either side of the central (*principal*) maximum. These are situated in directions in which the path difference BE is an odd multiple of $\lambda/2$, *i.e.,*

$$BE = a \sin\theta = (2n + 1)\lambda/2. \quad ...(b)$$

The intensity of these secondary maxima is much less and falls off rapidly as we move outwards.

The variation of intensity with distance from the centre of the central maximum. The intensities of illumination of the first and the second secondary maxima on either side of point O are 1/22 and 1/61 respectively, of the intensity of the central maximum. The rest of the secondary maxima are not easily visible as they decrease in intensity rapidly with the increase of obliquity.

In the slit is wide (*i.e., a* is large) θ is very small, and hence the central maximum is very narrow and there is a dose approximation to a perfect image. On the other hand, if the slit is narrow (*i.e., a* is small) θ is quite large and the central maximum is broad and flat. There will be no perfect image and the first minima lie far out.

Mathematical Investigation

Let us investigate in detail the intensity variation in the diffraction pattern of a single slit. The path difference *BE*, between the extreme rays is equal to *a* sin 6 which is equivalent to a phase difference

$$2a = \frac{2\pi a}{\lambda} \sin \theta. \qquad ...(c)$$

Now, suppose that the aperture is divided into a large number of strips, say *n*, of equal width parallel to its length, each strip forming an elementary source. Since the incident light is parallel, the amplitude of the wave from each strip may be taken to be the same but their initial phase will differ by a constant amount. The problem, therefore, reduces to the calculation of the resultant of a number of vibrations of equal amplitude and regularly increasing phases. By the method of vector analysis, the resultant is given by

$$R = \text{chord } OQ = 2r \sin \alpha,$$

where *r* is the radius of the circumscribing circle and 2α is the phase difference between the initial phase of the first and the last elementary waves. It is equal to ∠*QLX* and also equal to ∠*QCO*.

If s is the length of the are *OPQ*, then s must be directly proportional to the width *a* of the aperture, *i.e*, $s = n\ a = A_0$, say.

But $$s = 2r\alpha, \therefore 2r\alpha = A_0$$

or $$r = \frac{A_0}{2\alpha}, \text{ whence the chord}$$

$$OQ = 2r \sin \alpha = \frac{A_0 \sin \alpha}{\alpha},$$

and this denotes the amplitude of the resultant vibration so that the resultant intensity is given by

$$I = \frac{A_0^2 \sin^2 \alpha}{\alpha^2} \qquad \text{... (2)}$$

where $\alpha = \frac{\pi}{\lambda} a \sin \theta$ by equation *(c)* above.

Determination of Maxima and Minima

Since the magnitude of intensity at any point in the focal plane of the lens is a function of α and hence of θ it follows that a series of maximum and minimum values will be obtained.

Now the intensity is zero when $\sin \alpha$ = zero or $\alpha = \pm n\pi$, where n is an integer except the value n = zero and, hence $\alpha = 0$, when the intensity is a maximum as

$$\lim_{\alpha \to 0} \frac{A_0^2 \sin^2 \alpha}{a^2} = A_0^2 \qquad \text{...(3)}$$

Minima. The positions of the minima are, therefore, given by

$$\alpha = \pm n\pi$$

or $$\frac{\pi}{\lambda} a \sin \theta = \pm n\pi$$

or $$a \sin \theta = \pm n\lambda.$$

For the first minimum adjacent to central maximum

$$n = \pm 1$$

$\therefore$ $$\sin \theta = \pm \frac{\lambda}{a},$$

Other minima occur along directions obtained by giving n the integral values

$$\pm 2, \ \pm 3, \text{ etc.}$$

Let us examine the case when a = λ, sin θ = 1 ∵ θ = 90°, that is, light spreads out in all directions and the intensity decreases from the centre outwards giving us a broad central maximum but no other maximum is obtained.

Maxima. To determine the positions of the maxima of intensity in the diffraction pattern let us investigate the expression $\sin^2 \alpha/\alpha^2$. Put

$$Z = \frac{\sin^2 \alpha}{\alpha^2}.$$

Differentiating with respect to α, we get

$$\frac{dZ}{dx} = \frac{\alpha^2 . 2 \sin \alpha \cos \alpha - \sin^2 \alpha . 2\alpha}{\alpha^4}$$

$$= \frac{2 \sin \alpha \cos \alpha}{\alpha^2} - \frac{2 \sin^2 \alpha}{\alpha^2}.$$

To get a *stationary* value let us equate it to zero, then

$$\frac{\sin \alpha}{\alpha^2}\left(\cos \alpha - \frac{\sin \alpha}{\alpha}\right) = 0,$$

i.e., either sin α = 0, or tan α = α.

The first of these equations determines the position of the minima except that when α = 0, we have a maxima as explained above. The second equation determines the positions of the maxima and can be solved graphically by plotting the curves $y = \alpha$ and $y = \alpha \tan a$. The points of intersection of these two curves naturally give the roots of the equation tan α = α and hence the values of the maxima. The first is a straight line making an angle of 45° with the axes and the second a discontinuous curve of infinite number of branches with a symtotes at

$$\alpha = \pm \frac{\pi}{2}, \pm \frac{3\pi}{2}, \pm \frac{5\pi}{2} \ldots \ldots$$

and at intervals α = π.

By inspection of the points of intersection of the two curves we find that excepting the first root for which $\alpha = 0$, the values of a for α maximum value approach more and more to the value $(2n + 1)\pi/2$, where $n = 1, 2, 3$, etc.

So for all practical purposes maxima occur when

$$\alpha = \frac{3\pi}{2}, \frac{5\pi}{2}, \frac{7\pi}{2}$$

or more exactly, $= 0, 1.43\pi. 2.459\pi. 3.471\pi, \ldots \ldots$

Substituting these values of α in equation (2)*viz.*,

$$I = \frac{A_0^2 \sin^2 \alpha}{\alpha^2}, \text{ we get}$$

$$\underset{\alpha \to 0}{I_o} = A_0^2 \frac{\sin^2 \alpha}{\alpha^2} = A_0^2,$$

$$I_1 = \frac{A_0^2 \sin^2 (3\pi/2)}{(3\pi/2)^2} = \left(\frac{2}{3\pi}\right)^2 A_0^2$$

$$I_2 = \frac{A_0^2 \sin^2 (5\pi/2)}{(5\pi/2)^2}$$

$$= \left(\frac{2}{5\pi}\right)^2 A_0^2 \text{ and so on} \qquad \ldots(4)$$

Thus the intensities of the maxima decrease rapidly being approximately in the ratio of quantities,

$$1, \left(\frac{2}{3\pi}\right)^2, \left(\frac{2}{5\pi}\right)^2, \left(\frac{2}{7\pi}\right)^2 \ldots \ldots \qquad \ldots(5)$$

If the intensity of the first maximum is taken as unity, the values of the second, the third and the fourth will be approximately respectively.

$$\frac{1}{22}, \frac{1}{61}, \frac{1}{121} \text{ respectively.}$$

One half of the intensity variation curve, where the graph $y = \sin^2 \alpha/\alpha^2$ is plotted, its abscissa being angle α and its ordinates being the intensities. The first maximum is very much greater than the others and these again diminish very rapidly. For complete intensity pattern.

Thus the more complete investigation predicts not merely the positions of minima in the pattern but also the positions of maxima and their relative intensities.

Example. *In the Fraunhofer diffraction due to a narrow slit a screen is placed 2 meters away from the lens to obtain the pattern in its focal plane. Find the slit width if the first minima lie 5 mm on either side of the central maximum when plane waves of wavelength* 6×10^{-5} *cm are incident on the slit.*

For various minima to occur the path difference between the extreme rays is an integral multiple of λ, *i.e.*,

$$a \sin \theta = n\lambda,$$

where a is the width of the slit.

Here, $n = 1$; $\lambda = 5 \times 10^{-5}$ cm ; to obtain the value of sin θ, where x, the linear separation between the central maximum and the first minimum= 0.5 cm and $f = 200$ cm.

$$\therefore \qquad \sin \theta = \frac{0.5 \text{ cm}}{200 \text{ cm}}.$$

Hence

$$a = \frac{n\lambda}{\sin \theta} = \frac{1 \times 5 \times 10^{-5} \text{ cm}}{0.5/200}$$

$$= \frac{5 \times 10^{-5} \times 200 \text{ cm}}{5} = \mathbf{0.2 \text{ mm}.}$$

Fraunhofer diffraction at a circular aperture. The problem of diffraction of plane waves from a point source at an infinite distance passing through a narrow circular aperture is of great importance on account of its application to resolving power of telescopes and microscopes. It was first discussed by Sir G.B. Airy in 1835 and was solved with the help of well known Bessel's functions too elaborate to be considered here. According to this treatment, the diffraction pattern of a circular aperture consists of a central bright disc called Airy's disc surrounded by a system of dark and bright rings in the focal plane of the focusing lens *L*. These rings are known as Airy's rings. The intensity of light shades off gradually at the edges of the central disc and the rings and its distribution is similar to that already obtained in the case of a rectangular slit.

AB is a circular aperture of diameter *d*. Parallel light is incident on it and after diffraction it is brought to focus on a screen by a convex lens *L*. The intensity is maximum at the fecal point *O* of the lens where all the rays parallel to *CO* before incidence reinforce each other as they arrive there in the same phase. The rays reaching a point *P* after diffraction at an angle θ from the extreme ends *A* and *B* of the aperture have a path difference $BE = AB \sin \theta$. If *BE* is equal to one wavelength, then by the reasoning similar to that in the case of a single slit pattern, the *first minimum* occurs at *P* given by

$$AB \sin \theta = \lambda$$

For very small values of θ,

$$\sin \theta = \theta \frac{\lambda}{AB} = \frac{\lambda}{d},$$

where *d* is the diameter of the aperture *AB*. If *r* denotes the distance *OP*, that is, the radius of Airy's disc and *f* the focal length of the lens, then

$$\theta = \frac{OP}{OC} = \frac{r}{f}.$$

$$\therefore \quad \frac{r}{f} = \frac{\lambda}{d}$$

$$\text{or} \quad r = f \frac{\lambda}{d}.$$

In general, if path difference between the extreme diffracted rays is an integral multiple of A, there will be zero intensity at P and if it is an odd multiple of $\lambda/2$ there will be maximum intensity at P. The conditions for maxima and minima are given by

$$r_n = \frac{f}{d}(2n + 1)\frac{\lambda}{2} \text{ .... for maxima} \quad ..(d)$$

and $$r_n = 2n\frac{\lambda}{2} \text{ for minima.} \quad ...(e)$$

Since the circular aperture is symmetrical about the axis, the resultant diffraction pattern may be obtained by rotating about the same axis. The point P will trace out a circular ring of uniform intensity, minimum or maximum depending upon the conditions given in equations (*d*) and (*e*).

Thus the diffraction pattern is characterised by a central bright disc surrounded by alternate dark and bright concentric rings.

The exact calculation of the radii of the rings is very tedious and the mathematical analysis by Airy shows that for a circular aperture the "Polar Zones" are somewhat less effective, so that the radius of the first dark ring is slightly greater then that expected on the above calculations. Actually the first minimum occurs an at angle given by

$$\sin\theta = \frac{1.22\,\lambda}{d}$$

For small values of θ, $\sin\theta = \frac{1.22\,\lambda}{d}$. ...(6)

For a lens of focal length f, the radius of the first dark ring or the radius of the central disc of the pattern is given by

$$r = f\theta = f \times \frac{1.22\,\lambda}{d}. \quad ..(7)$$

Thus if *diameter d of the aperture is large, r the radius of the central disc is small, which is therefore well defined.*

Fraunhofer diffraction at two slits. The same graphic method may be applied to the case of two fine slits as for a single slit. Let *AB* and *CD* represent two rectangular slits each of width *a* and separated by an opaque interval of width *b* placed close and parallel to each other. Then the distance between the corresponding points of the two slits is equal.

Let a parallel beam of monochromatic light of wavelength A be incident upon this double slit and let each slit diffract the beam at an angle θ. The Path difference between the rays diffracted from the corresponding points *B* and C of the two slits is equal to

$$CE = (a + b) \sin \theta,$$

and the corresponding phase difference,

$$2\beta = \frac{2\pi}{\lambda} (a + b) \sin \theta$$

Half the phase difference, $\beta = \frac{\pi}{\lambda} (a + b) \sin \theta$ ...(*i*)

Here each slit can be treated as an independent small source of light, and if the light from the two slits be focused by a convex lens on a screen in the focal plane of the lens, the resultant pattern will be the same as that due to the interference from two sources of light placed at the corresponding points.

If 2α is the phase difference between the extreme rays from the first slit, then

$$2\alpha = \frac{2\pi}{\lambda} a \sin \theta \text{ or } \alpha = \frac{\pi}{\lambda} a \sin \theta. \quad ...(ii)$$

Analytically the resultant displacement y_1 due to the rays from the first slit is given by

$$y_1 = A \sin \omega t,$$

where $$A = A_0 \frac{\sin \alpha}{\alpha}. \quad ...(iii)$$

Here A_0 is the resultant amplitude of the direct rays, and A that of the rays diffracted at an angle θ from the first slit.

The resultant displacement y_2, due to the rays from the second slit is given by

$$y_2 = A \sin (\omega t + 2\beta) \qquad \text{....}(iv)$$

The resultant displacement y due to the rays from the two slits, diffracted at an angle θ is given by

$$\begin{aligned} y &= y + y_2 = A \sin \omega t + A \sin (\omega t + 2\beta) \\ &= A [\sin \omega t + \sin (\omega t + 2\beta)] \\ &= 2 A \cos \beta \sin (\omega t + \beta). \qquad \text{...}(v) \end{aligned}$$

The amplitude of the resultant

$$= 2A \cos \beta = 2A_0 \frac{\sin \alpha}{\alpha} \cos \beta \text{ by equation } (iii) \text{ above}$$

and the resultant intensity, I is proportional to $4\lambda_0 \frac{\sin^2}{\alpha^2}$ $\cos^2 \beta$. ...(7*a*)

Thus the resultant intensity in the diffraction pattern depends upon two variable factors :

(*i*) $A_0^2 \frac{\sin^2 \alpha}{\alpha^2}$, gives diffraction fringes similar to those due to a single slit, and

(*ii*) $\cos^2 \beta$ which gives a system of interference fringes due to the superposition of the wavelets from the corresponding parts of the two slits. The set of diffraction fringes has *greater* dispersion while the interference set of fringes which is superimposed on this and is visible only in the region of central maximum has *lesser* dispersion.

Nature of Diffraction Pattern

To know more about the nature of diffraction pattern let us make use of the fact that in practice θ is very small Then.

the amplitude, $A\left(=A_0\dfrac{\sin\alpha}{\alpha}, \text{ where } \alpha \text{ is } \dfrac{\pi}{\lambda}\theta\right)$ is maximum when $\alpha \to 0$, since $\lim\limits_{\alpha\to 0}\dfrac{\sin\alpha}{\alpha} = 1$, and it vanishes when $\alpha = \pm\,\pi, \pm\,2\pi, \pm\,n\pi$, so that for minima in the set of ' diffraction fringes, the corresponding values for θ are $\pm\,\lambda/1$, $\pm\,2\lambda/a$, $\pm\,3\lambda/a$, ... $\pm\,n\lambda/a$ and for subsidiary maxima α has values $\pm\,3\pi/2$, $\pm\,5\pi/2$, $\pm\,7\pi/2$, ... $(2n+1)\,\pi/2$. These results have already been discussed and, therefore, $\theta = \pm\,\dfrac{3\lambda}{2a}, \pm\dfrac{5\lambda}{2a}, \pm\dfrac{7\lambda}{2a},$ .... $\pm\dfrac{(2n+1)}{a}\cdot\dfrac{\lambda}{2}\cdot$

The variation of the intensity term $A_0^2\,\dfrac{\sin^2\alpha}{\alpha^2}$ as a function of θ.

Examining the second factor $\cos^2\beta$ for intensity variation, we

$$\therefore \frac{n}{\lambda}\,(a+b)\sin\theta = \pm\,n\pi, \text{ by equation } (i)$$

or
$$\sin\theta = \pm\,\frac{n\lambda}{a+b}$$

and since θ is small, we may write θ for $\sin\theta$,

$$\therefore \qquad \theta = 0 \pm \frac{\lambda}{(a+b)}, \pm\frac{2\lambda}{(a+b)}, \pm\frac{3\lambda}{(a+b)}, \ldots \qquad ...(vi)$$

When $\theta = 0$, we get the central maximum of the interference pattern along the direction in which the rays, are incident. This is also the direction of the central maximum determined by the first factor. Here all the rays arrive in phase,. Other values of

θ in integral multiples of λ give the maxima on either side of the central maximum called the principal maximum.

For minima,

$$\cos^2 \beta = 0 \text{ whence } \beta = \pm (2n + 1) \frac{\pi}{2},$$

where $n = 0, 1, 3,$ etc.

Substituting for β, we have

$$\frac{\pi}{\lambda} (a + b) \sin \theta = \pm \frac{(2n+1)\pi}{2},$$

or

$$\sin \theta = \pm \frac{(2n+1)\lambda}{2(a+b)}.$$

Writing θ for $\sin \theta$ as below :

$$\theta = \pm \frac{\lambda}{2(a+b)}, \pm \frac{3\lambda}{2(a+b)}, \pm \frac{5\lambda}{2(a+b)}, \ldots$$

Generalising the results, the minima of intensity occur along directions fur which the path difference between the two corresponding points on the slits is an odd multiple of $\lambda/2$.

The angular separation $\delta\theta$ of the interference set of fringes is therefore, $\lambda/(a + b)$. Since the intensity of illumination in the resultant diffraction pattern is given by

$$I = 4 A_0^2 \frac{\sin^2 \alpha}{\alpha^2} \cos^2 \beta, \qquad \ldots(8)$$

its variation with θ is obtained by the product of ordinates of only half of the diagrams. In this case the interference fringes form a much finer pattern superimposed on the relatively broad diffracted bands.

Next let us suppose that there are *three* slits instead of two, so we shall have to determine the resultant of three vectors of equal amplitudes and successively differing in phase by 2p. By an extension of the process outlined, we find that the resultant intensity comes to be

$$I = I_0 (1 + 2 \cos 2\beta)^{2*}. \qquad \ldots (9)$$

The maximum value of $(1 + 2\cos 2\beta)^2$ is 9 and the vertical scale should be increased by this factor. The maxima are much more intense and much narrower although they occur at the same angle as with two slits. Smaller subsidiary maxima also appear between them. By increasing the number of sifts the maxima become progressively more and more brighter and narrower and the intensity of subsidiary maxima rapidly falls. If the number of slits is very large the principal maxima become very sharp indeed and the subsidiary maxima are of negligible intensity. Such an optical device is called a *diffraction grating,* which we shall discuss in the following sections. But before we do so, we shall discuss an application of double slit.

Example. *A diffraction phenomenon is observed using a double slit with light of* $\lambda = 5000$ *A, slit width* $a = 0.02$ *mm, and the spacing between the two slits* $b = 0.10$ *mm. the distance of the screen from the slits is 1 meter. Calculate (i) the distant between the central maximum and the first minimum of the fringe envelope, and* (ft) *distance between any two consecutive double slit dark fringes.*

Here, $a = 0.002$ cm. 6=0.01 cm, and $\lambda = 5 \times 10^{-5}$ cm.

First minimum of the fringe envelope occurs at $\alpha = \pi$.

Since,
$$\alpha = \frac{\pi}{\lambda} \alpha \sin \theta$$

$$\therefore \quad \frac{\pi}{\lambda} = a \sin \theta = \pi$$

or
$$a \sin \theta = \lambda$$

Substituting, we get

$$\sin \theta = \frac{\lambda}{a} = \frac{5 \times 10^{-5} \text{ cm}}{0.002 \text{ cm}}$$

$$= 2.5 \times 10^{-2}$$

Since θ is small

$$\therefore \quad \sin \theta = \theta = 2.5 \times 10^{-2} \text{ radian.}$$

Since distance of screen from slits = 100 cm, the distance x of the not minimum from central maximum (which is at $\theta = 0$) is given by

$$\tan\theta = \theta \frac{x}{100} = 2.5 \times 10^{-2} \text{ cm.}$$

$$\therefore \quad x = 2.5 \text{ cm.}$$

(*ii*) Since $\beta = \frac{\pi}{\lambda}(a+b)\sin\theta$ and double slit dark fringes occur at

$$\beta = \pm(2n+1)\frac{\pi}{2}$$

$$\beta = \frac{\pi}{2}, \frac{3\pi}{2}, \frac{5\pi}{2} \ldots \ldots$$

Let θ_1 be the angle corresponding to the first dark fringe and θ_2 corresponding to second dark fringe.

$\therefore$ for first dark fringe,

$$\frac{\pi}{2} = \frac{\pi}{\lambda}(a+b)\sin\theta_1$$

$$\text{or} \quad (a+b)\sin\theta_1 = \frac{\pi}{\lambda}$$

$$\sin\theta_1 = \frac{5\times10^{-5}\text{ cm}}{2(0.002+0.01)\text{ cm}} = \frac{5\times10^{-5}}{0.024}$$

$$\text{or} \quad \theta_1 = \frac{5\times10^{-5}}{0.024} \quad \ldots(i)$$

For 2nd dark fringe,

$$\frac{3\pi}{2} = \frac{\pi}{\lambda}(a+b)\sin\theta_2$$

$$(a+b)\sin\theta_2 = \frac{3\lambda}{2}$$

$$\sin \theta_2 = \frac{3 \times 5 \times 10^{-5} \text{ cm}}{0.024 \text{ cm}}$$

or $$\theta_2 = \frac{3 \times 5 \times 10^{-5}}{0.024} \qquad ...(ii)$$

If x_1 and x_2 be respectively the distances of the 1st and 2nd minima from the centre,

then $$\theta_1 = \frac{x_1}{100} = \frac{5 \times 10^{-5}}{0.024}$$

and $$\theta_2 = \frac{x_2}{100} = \frac{3 \times 5 \times 10^{-5}}{0.024}$$

$\therefore$ the linear separation between first two minima,

$$x = x_2 - x_1 = \frac{3 \times 5 \times 10^{-5}}{0.024} \times 100 \text{ cm}$$

$$= \frac{5}{12} = \mathbf{0.42 \text{ cm.}}$$

Measurement of Diameters of Stars — Michelson's Stellar Interferometer

Principle. The measurement of the diameters, of stars is primarily based upon the nature of the interference pattern produced by a *double slit* mounted in front of a telescope objective.

Let S_1S_2 be a distant illuminated slit considered as source. Let , 2, 3, 4,..., n be the elements in the plane of the slit source parallel to the sides. The light from each of these elements, when incident on a *double slit* placed in front of a telescope will give rise to an interference pattern in the focal plane of the objective of equidistant maxima and minima superposed as shown graphically. The dotted curve at the top of the diagram (ft) indicates that unless the two slits have a suitable separation there will be maxima and minima in the composite pattern. If, however, the central maximum of nth element fails on the first maximum of the first element, the fringes will disappear completely and there will be uniform illumination. This is called *first order* disappearance. Theory shows that for the first order disappearance in the case of a single slit source, the angular width of the slit source is given by

$$\phi = \frac{\lambda}{D},$$

where D is the separation between the two slits in front of the telescope and λ the wavelength of the monochromatic source illuminating the slits. If instead, of a slit source, we have an illuminated circular aperture or a star, its angular diameter

$$\phi = \frac{1.22\,\lambda}{D} \qquad ...(10)$$

for a first disappearance. Hence smaller the value of ϕ the larger must be the value of D, the separation between the two slits.

Since fixed stars subtend a very small angle of the order of $\frac{1}{100}$ at the telescope objective, the measurements of the diameters of stars with objectives of small aperture's become quite difficult. For this purpose telescopes of large apertures are necessary. For example, if $\phi = \frac{1}{100}''$ $\lambda = 5.6 \times 10^{-5}$ cm, then for the first disappearance of the fringes the slit separation

$$D = \frac{1.22\,\lambda}{\phi} = \frac{1.22 \times 5.6 \times 10^{-5}\text{ cm}}{\frac{1}{100} \times \frac{1}{60} \times \frac{1}{60} \times \frac{\pi}{180}}$$

$$= \frac{1.22 \times 5.6 \times 10^{-5} \times 100 \times 60 \times 60 \times 180}{\pi}\text{ cm.}$$

$$= 1409\text{ cm}$$

$$= 14,09\text{ m},$$

which obviously requires a very large aperture.

Michelson's Stellar Interferometer

This difficulty was overcome by Michelson in a very igenious way by which the effective aperture D of the telescope objective. is considerably increased. The experimental arrangement consists of four plane fixed on a carefully constructed framework and attached to the telescope. The mirrors M_1 and M_2 are mounted on a girder along which they can be moved to change the distance D between them, to several meters Mirrors M_1 and M_2 are parallel to each other and so are M_2, and M_3 but fating each other.

The light from a distant star is incident on mirrors M_1 and M_2 and after reflection from M_3 and M_4 enters the telescope through the slits A_1 and A_2. If ϕ is the angular diameter of the star, the rays from the extreme edges of the star disc enter along full and dotted lines.

The interference bands are observed in the focal plane of the objective and their positions will depend upon the phase difference between the rays of light coming from the two ends of the diameter of the star. Since ϕ is very small, the path difference from M_1 to A_2 and from M_2 to A_2 are the same along the reflected paths of either ray. The optical path of one edge of the star to F, therefore, differs from the optical path of the other edge to F_1 by the amount $D \times \phi$ where D is the distance between the mirrors M_1 and M_2.

The outer mirrors M_1 and M_2 are moved symmetrically on the girders until the bands disappear. This corresponds to a variation in the path difference by 1.22λ, where λ is the average value of the wavelength of light from the star. Thus corresponding to the first disappearance of the fringes,

$$D \times \phi = 1.22\lambda$$

or

$$\phi = \frac{1.22\lambda}{D} \text{ (radians)},$$

which gives the angular diameter of the star. Using this result

ϕ can be measured by finding the smallest value of D which will cause the fringes to disappear.

Michelson mounted such an arrangement of mirrors in front of 2.54 m (= 100 inch) telescope at the Mount Wilson observatory and carried out his observations on the star. *Alpha Orionis.* Assuming the average value of $\lambda = 5750 \times 10^{-8}$ cm he obtained for $D = 3.065$ m a value of $\phi = 0.047''$ as the angular diameter of the star.

Combining this result with the distance of the star derived from its parallax, its linear diameter can be calculated.

It should be noted that it has been assumed that the slits A_1 and A_2 diffract equally in all directions. The fringe-spacing is characteristic of the distance d between the slits A_1, and A_2. Thus it may be said that this arrangement effectively gives a magnification of the optical path difference of D/d. This use of the relatively small apertures of telescopes also helps to reduce the effects of atmospheric in homogeneity which in the case of large aperture telescopes produces unsteadiness of the image.

By this arrangement the resolving power of the telescopes is increased to an extent limited only by mechanical considerations. For the application of the phenomenon of interference to the astronomical observations it is of the greatest importance.

Diffraction grating. The arrangement of a large number of equidistant narrow rectangular slits of equal 'width, placed side-by-side parallel to one another is known as a grating. Franahofer was the first to produce a grating in 1821 with, about 120 narrow slits to a cm by stretching fine wires over screws, and subsequently he succeeded in making a wiled grating over glass with a *fine* diamond point. He determined the wavelengths of all the principal Fraunhofer lines in solar spectrum with the help of his ruled gratings. In 1880 Rutherford ruled a grating on metal with the help of an automatic dividing engine. Rowland (1848 - 1901) wiled about 8000 lines to a cm and developed a technique of ruling them on concave mirrors also. The advent of concave grating dispensed with the use of a collimator and telescope. After long experimentation R.W. Wood found that a luminised Pyrex glass, is quite suitable for the construction of large gratings.

The diffraction grating therefore in its normal form consists of a well polished glass or metal Surface upon which large number of fine equidistant parallel lines varying from 12,000 to 30,000 to an inch are ruled with the help of a dividing engine. The ruled surface varies from two to six inches. Transmission gratings are plane and are ruled on glass surfaces. Reflection gratings may be plane or concave and are ruled on polished metal surfaces. Light reflected from spaces between the narrow rulings interferes to produce maxima and minima in exactly the same manner as the light transmitted by a transmission grating. Since these original ruled gratings are quite expensive, in the laboratories, their photographic reproductions or replicas of celluloid are generally used. A thin layer of a solution of a celluloid dissolved in a volatile solvent is poured over the surface of a grating ruled by a dividing engine and is allowed to dry. This plastic film when stripped from the grating retains a faithful impression of the rulings of the original grating and is mounted on a glass plate when a transmission grating is desired or on a silvered surface to form a reflection grating. Thus a number of replicas may be made from a single original and the recent rapid development in the plastic industry has made it possible to obtain very cheap gratings.

In all these cases the opaque portions of the grating are the roughened lines made by the ruling point and are called opacities. They do not allow any light to pass through them. While the clear spaces between the lines are known as transparencies which allow the light to pass through them. If the widths of a transparency and an opacity be a and b respectively, the distance $(a + b)$ is called a grating constant or grating element and is usually denoted by g, The points in the two adjoining transparencies separated by a distance $(a + b)$ are known as corresponding points.

Elementary Theory

Elementary theory of the plane transmission grating. The theory of the diffraction grating combines a problem in diffraction with one of interference,. Here each slit in the grating sends out a diffracted beam in which the intensity distribution is a function of the slit width and these diffracted beams then interfere with one another to produce the final pattern. *ABC* and *H* represents the section of a plane transmission grating, having a total number of *N* clear spaces placed perpendicular to the plane of the paper. Suppose a parallel beam of monochromatic light of wavelength λ is incident normally on the grating surface. Most of the light on issuing from the spaces goes straight but a part of it gets diffracted in various directions. The light after passing, through a convex lens *L* is made to converge on the screen in its focal plane forming dark and bright bands on both sides of the central maximum.

All the rays starting from *AH* reach *O* in phase with one another and so reinforce each other producing an image of the slits (central maximum). The rays diffracted at an angle θ with the grating normal reach P_1 on passing through the lens in different phases. Draw *AK* perpendicular to the direction of the diffracted light, then *CN* is the path difference between the rays diffracted from the two corresponding points *A* and *C* at an angle θ.

The path difference, $CN = AC \sin \theta = (a + b) \sin \theta$

If this path difference is an even multiple of $\lambda/2$, then point *P* be bright and if an odd multiple of $\lambda/2$, the point *P* will be dark.

Thus $(a + b) \sin \theta = \pm n\lambda/2$ for reinforcement

and$(a + b) \sin \theta = \pm (2n + 1) \lambda/2$ for, annulment,

where n is an integer, 0, 1, 2, 3, etc.

Thus the diffracted rays from any pair of corresponding points of the slits will produce constructive or destructive interference at a point P according as the path difference between them is an even or odd multiple of $\lambda/2$. This condition holds true for all the rays from the corresponding points of any pair of adjoining slits in the entire grating surface. We find, therefore, that brightness and darkness are alternate.

The point O where we get central bright band, has the value $\theta = 0$ corresponding to $n = 0$ so that there is no retardation and the light from all the slits arrives in the same phase.

This gives the central maximum at O.

For $n = \pm 1, \sin \theta = \pm \lambda/ (a + b)$.

This gives the direction of the first order principal maxima on either side of the central maximum, *i.e.*, at P_1 and P'.

For $n = \pm 2, \sin \theta = \pm 2\lambda/ (a + b)$.

This gives the direction of the second order principal maxima at P_2 sad P''on either side of the point O, beyond P_1 and P' respectively and so on, for the third, fourth and fifth order principal maxima on either side of the zero order maximum. This applies to the light of a definite wavelength. When white light is used rainbow- coloured bands appear at each of the points P_1 P', etc., the points of maximum intensity for red light being further removed from the centre than the points for violet light, since $\lambda_r > \lambda_v$. The equation

$$(a + b) \sin \theta = n\lambda,$$

where n = 0, 1, 2....... is the so-called grating law and shows that the directions in which the principal maxima appear depend only on the grating constant and the wavelength and not on the number of spaces in the grating.

It may be noted that the angles for the principal maxima are the same as those at which bright lines appear in the interference pattern, of two slits. The difference, however, lies not in the *positions* of the maxima but in the distribution of intensity between the maxima as will be seen. We find that as the number of slits is increased, the maxima become much brighter and much narrower.

This phenomenon can be easily seen by any one on looking at the filament of an incandescent lamp at a distance through a fine silken handkerchief. A central white filament with several coloured filaments on the sides will be seen.

In fact in a grating each order consists of principal maxima of different wavelengths inclined to each other and hence in different positions. In the case of white light, therefore each order consists of a spectrum and the grating may be said to form multiple spectra.

Secondary maxima and minima. Between any two consecutive principal maxima, there is a set of much smaller secondary maxima and minima. If for a small change in the angle of diffraction for any order maximum the path difference between the extreme rays, from the grating surface changes by wavelength λ, the change in the path difference between the rays from any pair of corresponding points of the two adjoining spaces will be λ/N, where N is the total number of lines in the grating. If we imagine the grating surface to be divided into two halves, then light rays issuing from any space in one half will differ in path by $\lambda/2$ from those issuing from the corresponding space in the other half and so they will mutually cancel out each other's effect. This gives the position of the first secondary minimum. If the path difference between the extreme rays changes by 2λ, $3\lambda(N - 1)\lambda$, *i.e.*, the change in the path difference between any pair of corresponding points of the two adjoining slits is $2\,\lambda/N,\ 3\lambda/N., \ldots \ldots\ (N-1)\ \lambda/N,$ there will be other minima in those directions. Thus in all there will be $(N-1)$ secondary minima alternated by $(N-2)$ secondary maxima between the two consecutive principal maxima.

Since this change of λ/N in the path difference between the rays from any pair of corresponding points of the two adjoining spaces required to produce the first secondary minimum dose to a principal maximum is very small, the principal maxima, are very sharp. The secondary maxima because of very small intensity are quite invisible and consequently between any two consecutive principal maxima given by the relation $(a + b) \sin \theta_{n}, = n\lambda$, there is practically uniform darkness.

Thus the first secondary minimum next *to* the nth principal maximum will occur displaced from this by an angle $\Delta\theta$ corresponding to a change λ/N in the path difference between the corresponding points, that is,

$$(a + b) \sin (\theta + \Delta \theta) = n\lambda + \frac{\lambda}{N}$$

Expanding and simplifying, we get

$$(a + b) \sin \theta + (a + b) \cos \theta\ \Delta \theta = n\lambda + \frac{\lambda}{N}$$

{ $\therefore$ $\Delta\theta$ is small $\sin \Delta\theta = \Delta \theta$ and $\cos \Delta\theta = 1$

Also by the grating law, $(a + b) \sin \theta = n\lambda$.

$$\therefore (a + b) \cos \theta\ \Delta \theta = \frac{\lambda}{N},$$

or
$$\Delta \theta = \frac{\lambda}{N(a+b)\cos\theta} = \frac{\lambda}{a'}.$$

where a' is the width of the projection of the grating width on a plane perpendicular to the direction of the given principal maximum.

More accurate theory of the grating. We have seen that when plane parallel monochromatic light falls on two slits, the phase difference between corresponding points is

$$2\beta = \frac{2\pi}{\lambda}(a + b)\sin\theta$$

and the resultant amplitude of vibrations from each slit is

$$A_0 \frac{\sin\alpha}{\alpha},$$

where $\alpha = \frac{\pi}{\lambda} a \sin\theta$. Extending the application of these results to N slits, the problem reduces to finding the resultant amplitude of N vibrations in a direction θ, each of amplitude $A_0 \frac{\sin\alpha}{\alpha}$ but the phase increasing in arithmetical progression, the common phase difference 2β being $\frac{2\pi}{\lambda}(a + b)\sin\theta$.

$$R = a\frac{\sin(n\delta/2)}{\sin(\delta/2)}.$$

In our problem,

$$a = A_0 \frac{\sin\alpha}{\alpha};$$

$$n\delta = N\frac{2\pi}{\lambda}(a + b)\sin\theta$$

and

$$d = \frac{2\pi}{\lambda}(a + b)\sin\theta$$

$$\therefore \quad R = \frac{A_0 \sin\alpha}{\alpha} \times \frac{\sin\left[N\frac{\pi}{\lambda}(a+b)\sin\theta\right]}{\sin\left[\frac{\pi}{\lambda}(a+b)\sin\theta\right]}$$

$$= A_0, \frac{\sin\alpha}{\alpha} \times \frac{\sin^2 N\beta}{\sin\beta},$$

where

$$\beta = \frac{\pi}{\lambda}(a + b)\sin\theta$$

Hence the intensity is given by

$$I = R^2$$

$$= A_0^2 \frac{\sin^2 \alpha}{a^2} \cdot \frac{\sin^2 N\beta}{\sin^2 \beta} \quad \text{...(12)}$$

The equation gives the distribution of intensity for a finite number of beams with equal amplitudes and phases increasing in arithmetical progression.

For $N = 1,$

$$I = A_0^2 \frac{\sin^2 \alpha}{\alpha^2},$$

which is equation (2)

For $N = 2,$

$$I = 4A_0^2 \frac{\sin^2 \alpha}{\alpha^2} \cos^2 \beta,$$

which is equation (8)

Here the intensity in the second focal plane of the lens which receives the diffracted beams is the product of two variable factors.

The first factor, $A_0^2 \frac{\sin^2 \alpha}{\alpha^2}$ gives the intensity distribution in the diffraction pattern of a single silt, and has been explained.

The other factor, $\frac{\sin^2 N\beta}{\sin^2 \beta}$ which may be supposed to represent the interference term, gives the distribution of intensity due to all the slits, which we shall now examine. Let it be denoted by Z. The intensities of maxima and minima will be obtained by equating its first derivative $dZ/d\beta$ to zero.

Now,

$$Z = \frac{\sin^2 N\beta}{\sin^2 \beta}$$

Differentiating with respect to β, we get

$$\frac{dZ}{d\beta} = \frac{\sin^2 \beta . 2N \sin N\beta \cos N\beta - 2\sin^2 N\beta \sin \beta \cos \beta}{\sin^4 \beta}$$

$$= 2N \sin N\beta \cos N\beta . \frac{1}{\sin^2 \beta} - 2 \sin^2 N\beta \left(\frac{1}{\sin^2 \beta} \cos\beta\right)$$

For a maximum or a minimum, $\frac{dZ}{d\beta} = 0$,

$$\therefore 2N \sin N\beta . \cos N\beta \frac{1}{\sin^2 \beta} - 2 \sin^2 N\beta \left(\frac{1}{\sin^2 \beta} \frac{\cos \beta}{\sin \beta}\right) = 0.$$

or $$2 \frac{\sin^2 N\beta}{\sin^2 \beta} N \frac{\cos^2 N\beta}{\sin^2 \beta} - 2 \frac{\sin^2 N\beta}{\sin^2 \beta} . \frac{\cos \beta}{\sin \beta} = 0$$

or $$2 \frac{\sin^2 N\beta}{\sin^2 \beta} (N \cot N\beta - \cot \beta) = 0 \qquad \dots (a)$$

Hence, either $\frac{\sin N\beta}{\sin \beta} = 0$,

or $N \cot N\beta - \cot \beta = 0$

Principal Maxima

Considering equation (*a*) when $\sin \beta = 0$, $\beta = \pm n\pi$, [$n = 1,2,3, \dots$], $N\beta/\sin \beta$, then becomes an indeterminate fraction 0/0. To find its value we employ the usual method of finding the quotient of the first derivative of both the numerator and the denominator. Then

$$\underset{\beta \to n\pi}{Lt} \frac{\sin N \beta}{\sin \beta} = \underset{\beta \to n\pi}{Lt} \frac{\frac{d}{d\beta}(\sin N \beta)}{\frac{d}{d\beta}(\sin \beta)}$$

$$= \underset{\beta \to n\pi}{Lt} \frac{N \cos N \beta}{\cos \beta} = \pm N.$$

Thus the resultant amplitude in these directions is proportional to N, *i.e.*,

$$R = A_0 \frac{\sin \alpha}{\alpha} N$$

and
$$I = R^2 = A_0^2 \frac{\sin^2 \alpha}{\alpha^2} N^2,$$

so that the intensities are proportional to N^2 and they are the maxima. These maxima are very intense and are called principal maxima as distinguished from secondary maxima described below. The positions of these principal maxima correspond to

$$\beta = \frac{\pi}{\lambda}(a + b) \sin \theta = \pm n\pi,$$

$$(a + b) \sin \theta = \pm n\lambda, \text{ where } n = 0, 1, 2,\dots \dots$$

This is the grating law for the principal maxima obtained already in discussing the elementary theory.

For $n = 0$, we get the central or zero order maximum. For $n = \pm 1, \pm 21, \pm 3$, etc., we get the first, second and third order principal maxima respectively on either side of the central maximum.

Minima and Secondary Maxima

A series of minima occur when $\sin N\beta = 0$, provided that $\sin P$ is not zero at the same time, for then $\sin N\beta/\sin \beta$ becomes zero. Thus for minima, we have

$$N\beta = \pm m\pi,$$

or
$$N \frac{\pi}{\lambda}(a + b) \sin \theta = \pm m\pi,$$

or
$$N (a + b) \sin \theta = \pm m\lambda,$$

where m has all the integral values except $0, N, 2N \dots nN$, because for these values of m, $\sin \beta$ also becomes zero (as $\beta = \pi, 2\pi, 3\pi, \dots$) and we have seen above that these positions correspond to principal maxima. There are obviously $(N-1)$ minima between two principal maxima.

Let us next consider equation (*b*), $N \cot N\beta - \cot \beta = 0$. Its roots other than those for which $\beta = n\pi$ (which correspond to principal maxima) give rise to another series of maxima termed secondary maxima much less intense than the principal maxima. These are $(N-2)$ between two principal maxima.

Intensity of Secondary Maxima

To find their intensity the fraction $\dfrac{\sin^2 N\beta}{\sin^2 \beta}$ has to be evaluated subject to the condition,

$$N \cot N\beta - \cot \beta = 0.$$

Now,
$$\frac{\sin^2 N\beta}{\sin^2 \beta} = \frac{1}{\operatorname{cosec}^2 N\beta} \cdot \frac{1}{\sin^2 \beta}$$

$$= \frac{1}{1+\cot^2 N\beta} \cdot \frac{1}{\sin^2 \beta}$$

$$= \frac{N^2}{N^2 + N^2 \cot^2 N\beta} \cdot \frac{1}{\sin^2 \beta}$$

$$= \frac{N^2}{N^2 + \cot^2 \beta} \cdot \frac{1}{\sin^2 \beta}$$

... [$\because$ $N \cot N\beta = \cot \beta$.

$$= \frac{N^2}{N^2 + \dfrac{\cos^2 \beta}{\sin^2 \beta}} \times \frac{1}{\sin^2 \beta}$$

$$= \frac{N^2}{N^2 \sin^2 \beta + \cos^2 \beta}$$

or $$\frac{\sin^2 N\beta}{\sin^2 \beta} = \frac{N^2}{1+\left(N^2-1\right)\sin^2 \beta}$$

which shows that the ratio of the intensity of these secondary maxima to the intensity of the principal maxima which varies as N^2 is

$$\frac{\dfrac{N^2}{1+\left(N^2-1\right) sin^2 \beta}}{N^2}$$

or $$\frac{1}{1+\left(N^2-\right)\sin^2 \beta}.$$

so that for large valuer of N, these secondary maxima are weak and invisible. But when N is small they may be observed. The principal maxima with four secondary maxima between a pair in a case with 6 slits. In general we have $(N - 1)$ minima and $(N - 2)$ secondary maxima between each pair of principal maxima.

Position of Secondary Maxima

The positions of the. Secondary maxima may be determined by expressing the equation

$$N \cot N\beta - \cot \beta = 0$$

in the form $N \tan \beta = \tan N\beta$ and finding the points of intersection of the curves,

(*i*) $y = N \tan \beta$ and

(*ii*) $y = \tan N\beta$,

in a manner similar to that used in the case of a single slit (Art. 15-4). The first equation indicates the curves *OM* and *NX* asymptotic to the line $\beta = \pi/2$ and the second represents the family of curves asymptotic to lines $N\beta = (2n + 1)\,\pi/2$. The curves in represent the case when $N = 6$.

The resultant intensity curve of the distraction pattern is

determined by the product of two factors, $\frac{\sin^2 \alpha}{\alpha^2}$ and $\frac{\sin^2 N\beta}{\sin^2\beta}$. Its value is obtained by multiplying the ordinates of the curve relative to a single aperture. The variations of the latter are very feeble as compared with the principal maxima so that the appearance is hardly affected.

Intensity and Width of Principal Maxima

(a) *Intensity.* The intensities of the principal maxima are given by

$$I = A_0^2 \frac{\sin^2 \alpha}{\alpha^3} \cdot \frac{\sin^2 N\beta}{\sin^2 \beta}.$$

Obviously, these are not equal because the expression $\frac{\sin^2 N\beta}{\sin^2 \beta}$ according to which the intensity is proportional to N^2, is only a part of the expression for the intensity ; it must be multiplied by $\frac{\sin^2 \alpha}{\alpha^2}$, a factor which corresponds to that in the case of diffraction at a single slit.

For the central maximum

$$I_0 = A_0^2 \left(\frac{\sin^2 \alpha}{\alpha^2}\right)_{\alpha \to 0} \cdot \left(\frac{\sin^2 N\beta}{\sin^2 \beta}\right)_{\beta \to 0}$$

$$= A_0^2 N^2 \qquad \text{...13}$$

The first principal maximum on either side of the central maximum occurs along the direction $\beta = \pm \pi$, where $\beta = \frac{\pi}{\lambda}(a + b) \sin \theta$ and along the same direction $\alpha = \frac{\pi}{\lambda}$ a $\sin \theta$.

If $(a + b) = 3a$, $\beta = \frac{\pi}{\lambda}(3a) \sin \theta = 3\left(\frac{\pi}{\lambda} a \sin \theta\right) = 3\alpha$

or $\alpha = \beta/3 = \pi/3$

for the first order principal maximum.

$$\therefore \quad I_1 = \left[\frac{\sin^2(\pi/3)}{(\pi/3)^2}\right]\left[\frac{\sin^2 N\beta}{\sin^2\beta}\right]_{\beta\to 0}$$

$$= A_0^2 N^2 \sin^2 60 \times \frac{9}{\pi^2}$$

$$= 0.68\ A_0^2 N^2 = 0.68\ I_0.$$

Similarly, $I_2 = 0.17\ I_0$.

Thus there is a rapid fall in the intensity of the principal maxima with increasing orders.

(b) *Width.* The angular width of principal maximum is the angle between the two first minima lying on either side of it. If θ_n is the angle of diffraction for a wavelength λ in the nth order, then

$$(a + b) \sin \theta n = n\lambda.$$

For the first minimum on either -side of a principal maximum a change in the phase difference between waves from the first and the last, slit is $\pm 2\pi$. The change in the phase difference between disturbances from the corresponding points of the first and the second slits or any two successive ones is, therefore $\pm 2\pi/N$, while the change in the path difference for these two disturbances is $\pm\lambda/N$.

Now for the nth maximum, the path difference between the corresponding points of two adjacent slits is $n\lambda$ and if $\propto$ be the angle which the first minimum makes with it, *i.e., one half of the angular width,* then

$$(a+b) \sin \theta n = n\lambda \quad ...(i)$$

and $(a + b) \sin (\theta n \pm \propto) = n\lambda \pm \lambda/N$ ...(*ii*)

Dividing (*ii*) by (*i*), we get $\dfrac{\sin(\theta_n \pm \alpha)}{\sin \theta_n} = 1 \pm 1/Nn.$

Expanding sin ($\theta n \pm \alpha$) and simplifying, we get

$$\sin \alpha = \frac{\tan \theta_n}{Nn} \text{ or } \alpha = \frac{1}{Nn \cot \theta_n}. \quad ...(14)$$

For large values of N, α must be very small as compared to θ_n.

For $N = 14000$, $n=1$, $\theta = 20°$ approx., the value of one half of angular width,

$$\alpha = \frac{\tan 20°}{14000 \times 1} \text{ radian}$$

$$= \frac{0.3640}{14000} \times \frac{180}{\pi} \times 60 \times 60'' = 5.3''.$$

Thus the principal maxima are so narrow and sharp that they appear as images of the slit of the collimator.

Absent Spectra. Sometimes it so happens that for certain values of the angles of diffraction 6 satisfying the relation

$$(a + d) \sin \theta = n\lambda,$$

there is no spectrum. This happens when for a given direction θ, the path difference between the diffracted rays from the two extreme edges of one space is equal to an even number of half wavelength (integral multiple of λ). For example, if this path difference is λ, the slit can be imagined to be divided into two halves and the path difference between any pair of corresponding points in the two halves will be $\lambda/2$, which will result in zero intensity in that direction. Thus the condition for a minimum for a single slit is given by

$$a \sin \theta = m\lambda, \quad ...(i)$$

where m is an integer which is not equal to zero. But the condition for the principal maxima in the nth order spectrum of a diffraction grating is given by

$$(a + b) \sin \theta = n\lambda. \quad ...(ii)$$

Thus when the conditions (*i*) and (*ii*) are simultaneously satisfied although the diffracted rays from all the transparencies of the grating *reinforce* each other *but* those from each transparent space mutually annul each other and hence the resultant intensity is zero, which means that the spectrum is absent.

From (*i*) and (*ii*), we have

$$\frac{a+b}{a} = \frac{n}{m}$$

or
$$\frac{b}{a} = \left(\frac{n}{m} - 1\right).$$

When the width of the opacity is equal to the width of the transparency,

$$\frac{b}{a} = 1$$

and
$$n = 2m\ ;$$

so that the second, the fourth, the sixth, etc. orders of the spectrum will be missing corresponding to the minima due to a single slit given by $m = 1, 2, 3$, etc.

Also when the width of the opacity is double the width of the transparency,

$$\frac{b}{a} = 2$$

and
$$n = 3m,$$

so that the third, the sixth, the ninth, etc. orders of the spectrum will be missing corresponding to the minima due to a single slit given by $m = 1, 2, 3$, etc.

Dispersive power of the grating. *The dispersive power of a grating is defined as the change in the angle of diffraction corresponding to a unit change in the wavelength of light used.* We know that the condition for the nth order principal maxima for a grating is given by

$$(a + b) \sin \theta = n\lambda.$$

This expression indicates that for the given order n and for the given value of the grating element $(a + b)$, the angle of diffraction changes with the wavelength. Thus if the wavelength changes from λ to $(\lambda + d\lambda)$, the corresponding angle of diffraction would change from θ to $(\theta + d\theta)$. The ratio $d\theta/d\lambda$ is called the dispersive power of the grating. Differentiating the above with respect to λ, we get

$$(a + b) \cos \theta \frac{d\theta}{d\lambda} = n,$$

or $$\frac{d\theta}{d\lambda} = \frac{n}{(a+b)} \times \frac{1}{\cos\theta} = \frac{n \times m}{\cos\theta}, \quad ...(15)$$

where $m = \dfrac{1}{a+b}$ = number of lines per cm length of the grating. Since for small values of θ, $\cos \theta$ is constant so that dispersion $d\theta$ is directly proportional to $d\theta$, that is, different spectral lines in any one order will vary in angle by amounts which are directly proportional to the difference in wavelength. Such a spectrum is called a normal spectrum. It is also clear from the expression that the dispersive power $d\theta/d\lambda$ *depends directly upon (i) m, the number of lines per unit length of the grating and (ii) n, the order of the spectrum.*

The equation above gives us the angular dispersion per unit wavelength in a given order. In photography of the spectra we often need the linear dispersion between the spectral lines. If f be the focal length of the lens employed to obtain the spectrum on the photographic plate and dl the separation between two wavelengths λ and $\lambda + d\lambda$, then $d\theta = dl/f$,

$\therefore$ the linear dispersion per unit wavelength is given by

$$\frac{dl}{d\lambda} = f \frac{d\theta}{d\lambda} = \frac{nf}{(a+b)\cos\theta} \quad ... (16)$$

i.e., linear separation between wavelengths λ and $(\lambda + d\lambda)$, than in the *n*th order is equal to

$$dl = \frac{nf}{(a + b)\cos\theta}\, d\lambda. \qquad \text{... (17)}$$

Normal Spectrum

We have seen that the spectrum formed by a grating is pure, *i.e.*, there is no overlapping or mixture of colours. This is not generally the case with prim spectra. There is another important advantage of the grating spectra. In a grating $g(= a + b)$ so that the spectrum formed by one grating is exactly similar to the formed by another grating. If g and g' be the constants of the two gratings. Then for a given wavelength the widths of the corresponding spectra are in the ratio $g' : g$ for the dispersion is inversely proportional to g', consequently the ratio of the widths occupied in the spectrum by any two colours is invariable. This is not the case with the spectra formed by different prisms, the relative dispersion of two colours being different with prisms made of different glasses. This defect in prism spectra is known as irrationality of dispersion. On account of the fact that all grating spectra are similar comparison can be easily made between two spectra studied at different places, and the diffraction spectrum is, therefore, called the standard of normal spectrum.

Spectrum formation. It is evident from the equation

$$(a + b)\sin\theta = n\lambda,$$

that for a particular order n, the light of different wavelengths will be diffracted in different directions. For light of longer wavelength the angle of diffraction θ is greater and for that of shorter wavelength θ is smaller, os that is instead of monochromatic light white light is used, each wavelength gives rise to its own principal maximum in each order. Thus we get a spectrum in all orders, red colour being in the outermost position and violet in the innermost position, except at the central maximum for which $d = 0$ and $\theta = 0$ for all wavelengths.

Determination of wavelength by diffraction grating. The diffraction grating provides us with the most accurate method for measuring the wavelength of light in any part of the spectrum in an absolute manner. The expression

$$(a + b) \sin \theta = n\lambda ,$$

indicates that this method requires the accurate determination of the angle of diffraction only, while the grating constant $(a + b)$ is given by the manufacturers. Since θ is large and its value can be determined, correctly, this method is much superior to the interference methods in which very small measurements are involved. For instance, in the prism method, $\lambda = \beta - \frac{f}{D}$ where both β and d are very small and cannot be measured with great accuracy.

We have also explained earlier that the principal maxima are much brighter and much narrower. This increased sharpness of the lines enables the angle θ to be measured with much greater precision, and hence makes possible a more accurate determination of wavelength.

For the determination of the wavelength of a given source of light by this method an ordinary spectrometer is employed. The collimator slit is illuminated first by a source of, say, sodium light and the telescope and the collimator are adjusted for parallel rays by, say, Schuster's method. To set the grating surface *normal* to the incident beam, the telescope and collimator are placed in a line so that the direct image of the slit falls on the cross-wires in the centre of the field of view The telescope is then turned round and fixed at 90° with this position. The grating is fixed *vertically* on the prism table so that the ruled surface lies at the centre of the table and is perpendicular to the line joining two of the levelling screws *a* and *b*. The prism table is rotated until an image of the slit reflected from the grating surface is formed on the cross- wires of the telescope. Prism table is adjusted suitably to get the image of the slit in proper position. The grating table is next turned through 45° in the proper direction and the plane of grating will now be the perpendicular to the axis of the collimator. The telescope is then rotated and the positions of the diffracted images, on either side of the central direct image are observed. To get well defined images the screw *c* of the table is rotated to set the rulings parallel 10 the slit. For a few other minor adjustments the reader may look up College Practical Physics book by the authors.

Sodium light is now replaced by the given source of light. If the light contains more than one wavelength we shall get the spectrum. The telescope is focussed on each diffracted image in turn on either side of the central image, and its positions read. This is done to eliminate any error on account of the grating not being set exactly normal to the incident beam. Let the angle between the two positions be taken as 2θ. Half of this angle gives the value of the angle of diffraction θ for that order.

Knowing $(a + b)$ the grating element, the order n of the spectra and the corresponding angle of diffraction θ , the wavelength λ of the light employed is calculated from the relation

$$\lambda = \frac{(a+b)\sin\theta}{n}$$

Note. If N is the number of lines per cm in the grating, then the grating element

$$(n)_{\max} = \frac{a+b}{\lambda}$$

Number of orders of spectra in a grating. The reader may be interested to know the *number* of orders that are visible in a grating with normal incidence. To find this maximum number consider the equation $(a + b) \sin 6=nA$ again. Now, the maximum value of $\sin \theta$ is 1, so that

$$(n)_{max} = \frac{a+b}{\lambda}.$$

If $(a + b)$ is between λ and 2λ only first order is seen and if between 2λ and 3λ first two orders are visible and so on.

Generally, speaking, $n < \frac{a+b}{\lambda}$. In the laboratory transmission gratings are frequently used which have about 14.500 lines to 2.54 cm. With such a grating we have for the first order and for $\lambda = 6 \times 10^{-5}$ cm.

$$\sin \theta_1 = \frac{6 \times 10^{-5} \text{ cm}}{2.54 \text{ cm}} \times 14500$$

$$= 0.3425 \text{ or } \theta_1 = 20°\text{-}2'.$$

For the second order $\theta_2 = 43°\text{- }15'$ and θ_3 is not possible as $\sin \theta_3 > 1$, so we observe two orders only on either side of the central maximum. It may be noted that a theoretical limit is reached when the grating element $(a + b) = \lambda$. Since $(a + b) \sin \theta = \lambda$ for the first order $\sin \theta = \lambda / (a + b)$. If $(a + b) < \lambda$, $\sin \theta$ becomes greater than I, which is impossible. Hence no spectrum is formed and the grating is rendered useless.

Example. *A parallel beam of monochromatic light is allowed to be incident normally on a plane transmission grating having 5000 lines I cm and the second order spectral line is found to be diffracted through 30°. Calculate the wavelength of the light.*

Here, $(a + b) = \frac{1}{5000}$ cm ; $n = 2$; $\theta = 30°$.

Using the relation

$$(a + b) \sin \theta = n\lambda, \text{ we have}$$

$$\frac{1}{5000} \text{ cm} \times \sin 30° = 2\lambda .$$

or $$\lambda = \frac{\sin 30° \text{ cm}}{10000} = \frac{0.5}{10000} = \mathbf{5 \times 10^{-5} \text{ cm.}}$$

Example. *Monochromatic light of wavelength 6.56×10^{-5} cm falls normally on a grating consisting of parallel wires equidistant from one another. The first order spectrum is observed at 15′ from zero position. Find the grating constant.*

From the relation $(a + b) \sin \theta = n \lambda$, we have

$$(a + b) = \frac{n\lambda}{\sin \theta},$$

where $(a + b)$ is the grating constant, n the order of the spectrum, λ the wavelength of light used and θ the angle of diffraction.

Here, $n = 1 ; \lambda = 6.56 \times 10^{-5} \text{ cm} ; \theta = 15'$.

$$\therefore \quad (a + b) = \frac{1 \times 6.56 \times 10^{-5} \text{ cm}}{\sin 15'} = \mathbf{0.0149 \text{ cm.}}$$

Example. *Calculate the angle between the central image of a lamp filament and its first diffracted image produced by a fabric with 160 threads per cm ($\lambda = 6 \times 10^{-5}$ cm).*

Here, $(a + b) = \frac{1}{160} \text{ cm} ; n = 1 ; \lambda = 6 \times 10^{-5} \text{ cm}.$

Using the relation, $(a + b) \sin \theta = \lambda$, we have

$$\sin \theta = \frac{1 \times 6 \times 10^{-5} \text{ cm}}{1 \text{ cm} / 160} = 160 \times 6 \times 10^{-5}$$

$$= 96 \times 10^{-4} = 0.0096$$

$$\therefore \quad \theta = \sin^{-1} (0.0096) = \mathbf{33 \text{ min.}}$$

Example. *Sodium light of wavelengths 5890 and 5896* A *is incident normally upon a plane transmission diffraction grating of 10,000 lines in 2.54 cm. Calculate the angular separation of the sodium lines in the spectrum of the first order as observed in the telescope of which the objective has a focal length of 24 cm and eye-piece 2 cm.*

Using the relation,

$$(a + b) \sin \theta = n\lambda, \text{ we have}$$

$$\sin \theta = \frac{n\lambda}{(a+b)} = \frac{\lambda}{(a+b)} \qquad \text{... [}\because n = 1$$

Differentiating, we get

$$\cos \theta \frac{d\theta}{d\lambda} = \frac{1}{(a+b)}.$$

$$\therefore \qquad \frac{d\theta}{d\lambda} = \frac{1}{(a+b)} \times \frac{1}{\cos \theta}$$

$$= \frac{1}{(a+b)} \times \frac{1}{\sqrt{1 - \frac{\lambda^2}{(a+b)^2}}}$$

$$= \frac{1}{\sqrt{(a+b)^2 - \lambda^2}},$$

or $$d\theta = \frac{d\lambda}{(a+b)} \text{ approx.}$$

Here, $(a + b) = 2.54 \text{ cm}/10{,}000 = 2.54 \times 10^{-4}$ cm

and $d\lambda = (5896 - 5890) \times 10^{-8} \text{ cm} = 6 \times 10^{-8}$ cm

$$\therefore \qquad d\theta = \frac{6 \times 10^{-8} \text{ cm}}{2.54 \times 10^{-4} \text{ cm}} = 3/12700 \text{ radian.}$$

If $d\theta_1$ be the angular separation of the *D* lines as observed through the telescope, then

$$\frac{d\theta_1}{d\theta} = \frac{F}{f}$$

= angular magnifying power of the telescope,

where *F* and *f* are the focal lengths of the objective and the eye-piece respectively.

$$\therefore \quad d\theta_1 = \frac{F}{f}\, d\theta = \frac{24 \text{ cm}}{2 \text{ cm}} \times \frac{3}{12700}$$

... [since F = 24 cm and f = 2 cm

$$= 0.0028 \text{ radian} = \mathbf{7.9'}.$$

which gives the required angular separation of the sodium lines.

Plane transmission grating in oblique incidence. If the incident light instead of falling normally on the grating surface meets it at angle i with the grating normal, and θ be the corresponding angle of diffraction then the path difference between the two interfering beams from the corresponding points A and C, is given by

$$MC + CN = AC \sin i + AC \sin \theta$$

$$= (AB + BC)(\sin i + \sin \theta)$$

$$= (a + b)(\sin i + \sin \theta).$$

where $AC = (a + b)$ = the grating constant (a being the width of transparency and b the width of opacity).

For the nth order spectrum

$$(a + b)(\sin i + \sin \theta) = n\lambda \qquad \text{... } (i)$$

Show minimum deviation in the same manner as those produced by a prism

The order spectrum is given by $D = i + \theta$

For the deviation to at be minimum,

$$dD = di + d\theta = 0$$

whence $\quad d\theta = di.$

Differentiating equation (i) and substituting the value of $d\theta = -\, di$, we have

$$(a + b)(\cos i \, di + \cos \theta \, d\theta) = 0,$$

or $\quad \cos i \, di - \cos \theta \, di = 0,$

therefore, $\cos i = \cos \theta,$

that is, $i = \theta.$

The path difference in case of the rays diffracted downwards, is given by

$$AN - MC = AC \sin \theta \; AC \sin i$$

$$= AC (\sin \theta - \sin i)$$

$$= (a + b) (\sin \theta - \sin i).$$

For the nth order spectrum

$$(a + b) (\sin \theta - \sin i) = n\lambda. \quad ...(ii)$$

Here the deviation produced is given by

$$D = \theta - i,$$

or $$dD = d\theta - di = 0,$$

for deviation to be minimum.

or $$d\theta = di$$

Differentiating (*ii*) and substituting the value of $d\theta = di$, we get

$$(a + b) (\cos \theta \, d\theta - \cos i \, di) = 0,$$

or $$\cos \theta \, di - \cos i \, di = 0,$$

whence $$i = \theta$$

Hence for obtaining minimum deviation which is essential for improving the definition of the diffracted image, incident and diffracted beams must be equally inclined with the grating normal.

For this position of the grating $i = \theta = \frac{1}{2} \; D_{min} = \frac{1}{2}\phi$, say, the expression for nth order spectrum, therefore, becomes

$$2 (a + b) \sin (\phi/2) = n\lambda. \quad ...(18)$$

The concave reflection grating. We have seen above that like the prism a plane transmission or reflection grating requires the use of lenses for collimating the incident light and focusing the diffracted spectra. To study the spectra in the ultraviolet region we cannot make use of glass lens as they absorb the ultra-violet light. For this region quartz lenses are employed. The use of lenses is accompanied by spherical and chromatic aberrations and also decrease in the intensity of illumination. Also since it is not possible to rule more than 10,000 lines to a cm the spectra produced by small wavelengths would be quite close to the central maximum. All these difficulties helped Rowland to make an important discovery of a concave reflection grating. First class reflecting gratings may be formed by ruling very fine parallel lines on a polished concave spherical surface usually of *speculum metal.* The ruling should be so spaced on the grating surface as to be *equidistant,* on the chord of circular arc joining the limits of the rulings so that such a grating satisfies the same equation as a plane grating. The lines act as obstacles and the spaces behave as polished strips. If a be the width of a space and b that of a ruling, then $(a + b)$ is the grating constant and two points on the adjoining spaces separated by a distance $(a + b)$ are called corresponding points as in a transmission grating. It is also important to note that the surface of the grating should be highly reflecting.

Since the surface of the grating is concave spherical, the diffracted images are *automatically* focused without the use of lenses. The concave reflection gratings are free from all the drawbacks of plane gratings and the use of lenses. They can be conveniently used for the study of any region of the spectra.

Rowland deduced a remarkable property of a concave reflection grating that if a narrow slit source parallel to the rulings is placed on the circumference of a circle of which the diameter is equal to the radius of curvature of the grating surface, the various spectra of the source are automatically focused at different positions on the circumference of the same circle when the grating is so placed that-this circle is tangent to it at its mid-point. Such a circle is the focal carve of the concave grating and is known as Rowland circle.

Focal carve of a concave grating—Rowland circle. Let us assume that AC (= $a + b$), is a single element of a concave grating G_1G_2, with lines perpendicular to the plane of the paper and S is a source of light. Let, us consider two rays SAP and SCP incident at angles i and $i + \delta i$ and diffracted at $\theta + \delta\theta$ respectively and let the normate to the curved grating surface at A and C meet at its centre of curvature C_1. Denote the small angles at S, C_1 and P by. α, β and γ respectively. Then clearly, we have

$$\left.\begin{aligned}\alpha + i &= i + \delta i + \beta \\ \beta + \theta &= \theta + \delta\theta + \gamma\end{aligned}\right\} \because \text{ they are supplement of equal angels.}$$

Consequently, $di = \alpha - \beta$ and $d\theta = \beta - \gamma$.

For P to be on a spectral line corresponding to light of wavelength λ and of order n, the grating equation

$(a + b)(\sin i - \sin \theta) = n\lambda$,

must hold for images to lie on opposite sides of the normal. If light is homogeneous

$$\sin i - \sin \theta = \text{constant}$$

or $\cos i \, di - \cos \theta \, d\theta = 0$.

Hence substituting for di and $d\theta$ from above we obtain

$$(\alpha - \beta) \cos i - (\beta - \gamma) \cos \theta = 0. \qquad ...(i)$$

Let $r_1 R$ and r_2, be SA, $C_1 A$ (= C_1C) and PA respectively,

$$\text{Then } \alpha = \frac{(a+b)\cos i}{r^1}, \quad \beta = \frac{a+b}{R},$$

$$\Upsilon = \frac{(a+b)\cos\theta}{r^2}$$

Therefore (i) becomes

$$(a + b)\left[\frac{\cos i}{r_1} - \frac{1}{R}\right]\cos i - (a + b)\left[\frac{1}{R} - \frac{\cos\theta}{r_2}\right]\cos\theta = 0$$

or $$\frac{\cos^2 i}{r^1} - \frac{\cos i}{R} - \frac{\cos\theta}{R} + \frac{\cos^2\theta}{r^2} = 0,$$

whence $$r_2 = \frac{r_1 R\cos^2\theta}{r_1(\cos i + \cos\theta) - R\cos^2 i}.$$

Now if S moves along the curve $r_1 = R\cos i$ [a circle of diameter R], then

$$r_2 = \frac{R^2 R\cos^2\theta\cos i}{R\cos i[\cos i + \cos\theta] - R\cos^2 i} = R\cos\theta \quad \text{... (19)}$$

i.e., P also moves along the circle, whose diameter is R. The circle described by P is the focal carve, of the grating and is called the Rowland circle. Hence *if the grating curve be a circle of radius R, a Source situated on a circle described on R as diameter will give spectra situated on the same circle.*

It is this important property which has been utilized in a masterly manner by Professor Rowland in the construction of his concave gratings. This property has been utilized in the case of various types of mountings of concave reflection gratings.

Elementary theory of concave reflection grating. Let G_1G_2, be the section of concave reflection grating with lines perpendicular to the plane of the paper. A and Care the corresponding points in consecutive polished spaces so that their distance apart is $(a + b)$, the grating element, and C, is the centre of curvature of the grating. S is a fine slit lying on the Rowlond circle and held parallel to the lines of the grating. SA and SC are the two rays incident at an angle i at the corresponding points A add C, C_1A and C_1C being normals at

A and C respectively. The diffracted rays AP and CP make an angle θ with the normals and meet at P on the circumference of the same circle to give a spectrum. The angles of incidence are equal because they stand on the same arc SC_1. Similarly, the angles of diffraction are equal because they stand on the same arc PC_1.

To find the path Difference between the rays starting from S and reaching P, draw AH and CAT normals to CS and AP respectively.

The path difference

$$= (SC + CP) - (SA + AP)$$

$$= (SH + HC + CP) - (SA + AK + KP)$$

$$= HC - AK$$

...[since $SH = SA$ and $KP = CP$

$$= AC \sin i - AC \sin \theta)$$

$$= AC (\sin i - \sin \theta)$$

$$= (a + b) (\sin i - \sin \theta).$$

To get the condition for the nth order spectrum to lie at P using monochromatic light of wavelength λ, the path difference is given by

$$(a + b) (\sin i - \sin \theta) = n\lambda. \quad \text{... (20)}$$

Since AC may be any grating element, whatever, it follows that if diffracted rays from any two consecutive grating spaces reinforce each other at P, those from all pairs of grating spaces will do so.

Concave Grating Mountings

There are several mountings of concave gratings adapted for different purposes :

(1) *The Rowland mounting.* This mounting in addition to making use of the property of the Rowland circle mentioned above enables the observations on the spectrum to be made in a direction normal to the grating so that the *dispersion is linear giving us a normal spectrum.* The grating G, and the photographic camera P are fixed at the opposite ends of an iron girder GP whose length is equal to tie radius of curvature of the grating (diameter of the Rowland's circle which is more than 12 meters). The ends of this girder rest on two carriages which are free to rotate about and move along two tracks mutually at right angles to each other. The slit is placed at C, the intersection of the two tracks. In this way the condition that the slit and the centre of the grating lie on the circumference of a circle whose diameter is equal to the radius of the grating surface is satisfied. Since in addition the plate is also on the circumference of this circle, the spectra are always in focus. The angle of incidence can be varied by sliding the girder R, the angles of diffraction being always zero, since P is always at the centre of curvature of the grating. The movement of G enables the plate to cover the different orders of spectra. Since in the expression,

$$(a + b)(\sin i - \sin \theta) = n\lambda, \text{ angle } \theta \text{ is zero,}$$

$$\therefore \quad (a + b) \sin i = n\lambda,$$

or

$$(a + b) = \frac{PC}{PG} = n\lambda,$$

or

$$\lambda = \frac{a+b}{n \times PG} \times PC = \left(\frac{a+b}{nR}\right) PC \qquad ...(21)$$

where

$$R = (PG)$$

is the diameter of the Rowland circle.

Thus the length PC is proportional to λ giving us *normal* spectra, and can be calibrated to give the wavelengths directly. When the girder *R* is moved the grating and the photographic plate move with it and thus the spectra in different orders may be photographed.

The additional advantage in this mounting is that the initial direction of the beam remains fixed. This mounting is useful in studying the *solar Spectrum*. Rowland carried out all his celebrated researches on the solar spectrum with this form of mounting. The drawbacks in this mounting are that (*i*) it shows astigmatism particularly with higher order spectra thus causing diminution in brightness of lines, and (*ii*) it not only requires a large space frequently prohibitive but alto a room which can be darkened completely.

(2) *The Paschen mounting*. This form of mounting is very commonly used for *large* concave gratings. The grating and the slit are arranged in fixed positions on the Rowland circle, the slit being exactly parallel to the rulings on the grating. The light is made incident on the grating at any desired angle depending upon the problem under investigation and various order diffraction spectra will be in focus on the circle. The photographic plates are held in a *semi-circular* steel frame which bends them by means of clamps so as to coincide with the *focal curve*.

Its chief advantage therefore is that it enables any portion of the spectrum to be photographed in any of the available order at the same time without *altering* the adjustments. The ranges of the visible spectrum in the first three orders, C being the position of the central image. No other mounting covers such a wide spectral range under similar conditions. Its disadvantages, however, are that (*i*) there is a high astigmatism in every order except the zero order, (*ii*) the spectrum is not normal so that the dispersion is *non-linear* except when the rays fall back on the camera normally. But this defect is far out-weighed by the advantage of coveting a wide spectral range.

(3) *The Eagle mounting.* The light under examination after passing through a slit *S is* totally and internally reflected by a prism *Q* and then becomes incident on the grating *G*. The grating is set so at to utilize that diffracted light, which is returned directly back along the same path as the incident beam or nearly so and is focused on the photographic plate *P* by a slight tilt of the grating, the whole, apparatus is enclosed in a light-tight double-walled box lagged with wool. To secure a sharp image of all the spectral lines it is necessary that the slit and the photographic plate *P* should both be situated on the circle described on the radius of curvature of the grating at diameter, viz., on the Rowland circle. Therefore, the slit is mounted on the side of the box in a tube in such a position that its virtual image after total reflection in the prism is formed at a point just below the centre of *P*. As this arrangement requires a considerable variation of the distance between the grating and the slit for different angles of incidence needed to pass from one spectral range to another the grating G must be capable of to and fro motion along a straight track and also of rotation about a vertical axis to let the Rowland circle pass through *P* The plate holder is also capable of rotation about a vertical axis through its centre. This mounting has the advantage of being cheap and quite compact requiring less space, enables temperature controls to be easily affected even with long exposures and has therefore, largely replaced the Rowland and Paschen forms. It is commonly used in vacuum petrography for the investigation of the ultra-violet region below 2000 Å.

Since air absorbs these wavelengths it must be pumped out of the pectrograph. Another advantage lies in the fact that astigmatism is least in this type of mounting.

Example. A *Rowland grating has 15,000 lines in 2.54 cm. Calculate the angular separation of the two mercury lines 5770 Å, and 5451 Å in the spectrum of first order.*

The relation for the concave grating is $(a + b)(\sin i - \sin \theta) = n\lambda$.

To find the angular separation between the two lines of wavelengths λ and $(\theta + d\lambda)$ we differentiate the above equation and get

$$-(a + b) \cos \theta \frac{d\theta}{d\lambda} = n,$$

$$\text{or } \textit{numerically } d\theta = \frac{n d\lambda}{(a+b)} \times \frac{1}{\cos \theta}.$$

$$\text{Here, } \quad d\lambda = (5770 - 5461) \times 10^{-8} \text{ cm}$$

$$= 309 \times 10^{-8} \text{ cm} ; n = 1 ;$$

$$(a + b) = \frac{2.54}{15{,}000} \text{ cm}$$

and for Rowland grating, $\theta = 0$.

Substituting these values in relation (*i*), we get

$$d\theta = 1 \times 309 \times 10^{-8} \text{ cm} \times \frac{15{,}000}{2.54 \text{ cm}}$$

$$= 0.0182 \text{ rad.} = \mathbf{1^\circ\text{-}3'}.$$

Comparison of grating and prism spectra. The following are the points of difference between the two types of spectra:

1. A grating gives a number of spectra on either side of the central maximum while a prism gives only one spectrum.

2. The intensities of the grating spectra are less than the intensity of the prism spectrum. For, in the case of a prism the whole energy of the incident light is confined to a single spectrum while in the case of a grating this energy is distributed amongst different order spectra, the central undeviated bright band where no spectrum is formed, accounts for the greatest part of the energy.

3. The grating spectra are much purer than the prism spectrum which is practically impure.

4. In the grating spectra the deviation produced increases with the increase in the wavelength of the light used. Therefore, it *is more for red light than for violet but* it is quite independent of the material of the grating. While in the prism spectrum the deviation increases with the angle of the prism and being also dependent on the refractive index of its material it decreases with the increase in the wavelength *of* light used. *It is more for violet light than for red.*

5. The dispersive power

$$\left[\frac{d\theta}{d\lambda} = \frac{n}{(a+b)\cos\theta}\right]$$

in the grating spectra increases with the order of the spectrum. It is *almost constant* for a particular order spectrum, for the variation in θ is usually small. This means that in any one order, the change in the angle of diffraction is *directly* proportional to the change in wavelength, that is, *the spectrum is equally drawn out in the entire range.* While in case of a prism spectrum the dispersive power

$$\frac{d\mu}{\mu - 1} \qquad \text{...cf. Eqn. (1)}$$

being dependent on refractive index does not remain constant but increases with diminishing wavelength. *A prismatic spectrum, therefore, is more drawn out towards the violet end.* The relative dispersions of any two wavelengths are also different in the two prisms. For these reasons the prism spectra are said to be irrational whereas grating spectra are rational.

6. For a given source of light the spectra produced by one grating are exactly similar to those produced by another grating except that the spectrum length is directly proportional to the number of lines per cm of grating, that is, by doubling the number of lines per cm of the grating, the width of any part of the spectrum is doubled, whereas the spectra produced by two prisms of different material are never similar being dependent on the chromatic dispersion $\mu d/d\lambda$.

7. The resolving power of a grating is independent of the grating element and the wavelength of light used but depends upon the order of spectra and the total number of lines in the grating so that with a commonly used grating having 6000 lines, to a cm, the resolving power in the first order is $n\,N = 1 \times 6000 = 6000$. The resolving power of a prism depends upon the length of the base, refractive index of its material and the wavelength of light used. Its magnitude $\left(t\dfrac{d\mu}{d\lambda}\right)$ for a flint glass prism of 1 cm base is barely $1 \times 800 = 800$, which is much smaller than that of a diffraction grating of 1 cm width.

Scattering of light by small particles—Blue of the sky. This is a process involving the interaction of radiation and matter. When a beam of white light is made to pass through a medium consisting of small particles, say, a water tank with a few drops of milk, it is observed that the light seen in a direction at right angles to the path of the beam has a distinctly bluish colour while the transmitted beam has a corresponding excess of the less refrangible wavelengths at the red end of the spectrum. This phenomenon is known as scattering of light and occurs when the size of the particles of the medium is very small as compared to the wavelength of incident light. Tyndall and Rayleigh studied this phenomenon in detail and the latter showed by the method of dimensions that the amplitude a, of the scattered light should vary *inversely* as the square of the wavelength λ. As the intensity of light varies as a^2, we have a once that

$$I \propto \frac{1}{\lambda^4}$$

that is, the intensity of scattered light varies inversely as the fourth power of wavelength. This is known as Rayleigh's law of scattering. Smaller is the wavelength, therefore, greater is the scattering. Blue light, whose wavelength is half that of red light will be 16 times more intensely scattered than the red light Rayleigh actually measured the relative share of each wavelength in the sky light and found it in close agreement with his fourth power law.

The blue colour of sky can thus be easily accounted for. The light of the sun passing through the atmosphere is scattered by the dust particles, minute water particles and even the molecules of air. When one views the sky at a distance from the sun it is this scattered light which is rich in shorter wavelengths, *viz.*, blue colour that reaches the observer's eye. The question is, what becomes of the violet light which is shortest in wavelength ? The scattered light which contains an excess of green, blue and violet is scattered again so that the excess of shortest wavelength is removed on the way to observer's eye. The resultant light which finally reaches him has a maximum intensity in blue region. If it were not for our atmosphere, the sky would be perfectly black. Also it is why the light reaching us *directly* from the sun and the surrounding sky at sunrise and sunset looks red being deprived of the shorter components on the way. During these hours the light has to pass a much greater distance of the atmosphere containing fine dust and smoke particles so that the blue light is removed from the direct beam more effectively than the red, the result being that sky is deep blue while the transmitted light being rich in longer components makes the sun appear red. The same phenomenon accounts for the blue colour of the smoke from a cigarette and the greyish colour of many minerals.

These appearances can be easily imitated by passing a parallel beam of white light through a long glass tank containing sodium thiosulphate. When a little dilute sulphuric acid is added, particles of sulphur begin to be formed which gradually grow in size. Scattered blue light will indicate the path of the beam of light through the tank. The transmitted light will be observed to change first to yellow and then, as sulphur particles grow bigger, it becomes more and more red and finally it is practically cut off. As this occurs the scattered light becomes gray.

Halos and Coronas

There are many types Of halo phenomena. Some are quite uncommon in our latitudes and are observed, unlike rainbows, for all altitudes of the sun. They are the so called *halos* and are due to the refraction of the sun or the moon light through small ice crystals present in the air.

They are generally coloured circles sometimes bigger and some-limes smaller, surrounding the sun or the moon with red on -the in- side and white on the outside, the angular diameter of the smallest being about 45°. The outer circles are fainter than the inner ones. Sometimes it is an ellipse circumscribing the inner circular halo.

However, there are other types of halos known as *coronas* which are quite common in our latitudes and are seen encircling the sun and the moon. They are caused by diffraction of light by small drops of water hanging in the air during a mist or haze. By diffraction coloured circles are observed, blue lying on the inside and red on the outside. If the drops upon which light falls are of the same size in the different pans of the sky, the angles of diffraction for the same colour will be the same and the coronas will be perfect circles ; while if the drops are of different size the angle of diffraction for the same colour will vary as a result of which the coronas will-not be perfect circles. Thus the size and shape of the coronas give us an idea of the size of the droplets hanging in the air. The smaller the drops the larger in size the coronas but they are generally close to the source of light and possess small diameters.

If a strong source of light, say, an electric lamp is placed behind a small circular hole in a sheet of metal or card board and a glass plate lightly dusted with lycopodium powder is held in a stand between the observer and the circular hole the imitated diffraction coronas can be easily observed concentric with the hole. The diameters of these coronas may be measured to determine the size of the particles of the lycopodium powder. With monochromatic light a series of bright and dark rings are seen. According to Airy's calculation the first dark ring is seen in a direction making an angle θ_1 with the incident light, where $\theta_1 = 1\ 22\lambda/d$, where λ is the wavelength of light and d the diameter of the particle. The second and third dark rings are seen in directions $\theta_2 = 2.27\lambda/d$ and $\theta_3 = 3\ .23\lambda/d$. Let the *distance* between the plate and the hole be D_1 and radius of the first dark ring r_1, so that the direction in which the first ring is seen is $\theta_1 = r_1/D_1$, the angle being small. Then $\frac{r_1}{D_1} = \frac{1.22\lambda}{d}$, whence knowing r_1 D_1 λ, d may be calculated (see some advanced Practical Physics book for formula and practical details).

4

Interference Phenomenon

Some very useful and interesting applications of the interference phenomenon are provided by the use of some instruments known as Interferometers. They are the instruments which employ large path differences between the two interfering beams to obtain interference. They are used to study the fine structure of spectral lines and the variation of refractive indices of glass under different conditions of temperature and pressure, and also of solutions with change of temperature and density. The interferometers employed for the study of refractive indices are often called refractometers.

Michelson Interferometers

It is an instrument in which light from an extended source is divided into two parts by partial reflection and transmission (division of amplitude). These beams are sent in two directions at right angles to each other and are again brought together after reflection from plane mirrors to form interference fringes which are observed and examined with the help of a telescope.

It consists of two *optically* worked and *equally* thick plane parallel plates of glass A and B cut from the *same* piece. Plate A is partially silvered on the back surface so that parallel beam of incident light from a monochromatic source S is divided into a reflected and a transmitted beam of equal intensity. M_1 and M_2 are two highly polished *front-silvered* plane mirrors. The mirror M_1, is mounted on a carriage C and can be moved forward or backward slowly with the help of a micro-meter screw exactly parallel to itself along well-machined-tracks. This *slow* and *accurately* controlled motion of the mirror M_1 is accomplished by means of steel screw E of uniform pitch and fitted with a graduated drum D capable of reading a displacement of the order of 10^{-3} cm. The two mirrors M_1 and M_2 are also provided at the back with levelling screws for changing their planes according to need.

The function of the lens L is to make the source of light an extended one. The plates A and B are held parallel to each other and at 45° with the mirrors M_1 and M_2 which are made perpendicular to each other. Let a ray of light starting from S be incident on the plate A. It is partly reflected from its back-silvered surface and goes towards M_1 and partly transmitted towards M_2. These two split rays after reflection from these mirrors are returned to A so that the ray sent back by M_1 is transmitted through that plate while the ray sent back by M_2 is reflected from it and these two rays after reuniting at O travel in the direction OT and produce interference, since these resultant rays are derived from the same single ray SO. The phase changes on reflection at M_1 and M_2 are similar and Edser found that when the silver coating of A is thin, the phase changes on reflection from opposite sides of it in air and in glass respectively are each equal to π. Thus when the paths OM_1 and OM_2 are equal or differ by an integral multiple of A, the resulting rays will be in the same phase and will reinforce each other. When the paths differ by Odd number of $\lambda/2$, destructive interference will occur. The resulting interference pattern may be observed through a telescope T or the naked eye It may be noted that although the geometrical paths A to M_1 and A to M_2 may be equal yet they are not the corresponding optical path lengths because the split ray travelling towards M_1 traverses the glass plate A twice whereas the path of the ray towards M_2 in the absence of the plate B lies entirely in the air. It is to render the paths of the two rays symmetrical that the compensating glass plate B of equal thickness with A is introduced in the path of the transmitted beam. This makes the transmitted beam travel through B twice. Thus both the beams, one reflected from M_1 and the other from M_2, reach the point O for interference after travelling two equal thicknesses of glass. To obtain fringes the two mirrors are set *exactly* perpendicular to each other by means of screws shown on M_2. The interference pattern in general consists of *concentric circular* fringes.

A marked difference in this case from some of the foregoing methods is the production of interference between rays differing in path by very many wavelengths. In the case of Fresnels's bi-prism and the Newton's rings, for instance, the path difference amounts to a few wavelengths only.

Adjustments. In order to obtain the fringes some important requirements should be met.

Light must come from an *extended* source and it should be nearly monochromatic. A large sodium flame or a mercury vopour lamp can serve the purpose. If the source is small a lens L can be helpful for extending the source.

The distances of the mirrors M_1 and M_2 from the back surface of the plate A must be almost equal to within a millimeter. This can be done with the help of a millimeter scale.

The mirror M_2 is now made perpendicular to the mirror M_1. To do this a pin is placed between the source and the plate A. On looking along TO two pairs of images of the pin will be seen, one due to reflection at the front surface of A and other by reflection at its back surface. The adjustment of the plane of the mirror M_2 with the help of tilting screws at its back is made till one pair of images coincides exactly with the other pair when the interference fringes should appear. The pin is then removed. If now the two optical paths are exactly equal the field of view will be totally dark. A slight change in the path by moving the mirror M_1 *parallel* to itself will produce circular interference fringes. By further adjustment when the mirror M_2 is exactly perpendicular to M_1 a set of concentric circular fringes will be obtained. They will be situated at infinity and can only be seen if the eye or the telescope is focused for infinity.

Since the transmitted ray on its return journey suffers reflection at O on the plate A while travelling from a rarer-to-a-denser medium, a phase difference of π or a path difference of $\lambda/2$ is introduced, if the back surface of the plate is unsilvered. If this surface is thinly silvered then the ray travels from air to silver and silver to glass thus introducing a phase change of 2π, with the result that the central fringe becomes bright.

Circular fringes in Michelson's interferometer. The fringes produced by Michelson's interferometer are of various forms depending upon the separation, d, between M_1 and M'_2 (image of M_2) and the angle, ϕ, between their surfaces. Two cases of practical interest arise when (*i*) $\phi = 0$ and $d \neq 0$ and (*ii*) $d = 0$ and $\phi \neq 0$. In the first case concentric circular fringes are produced while in the latter case the central fringe is straight and all others convex towards it.

The concentric circular fringes are the most important type of fringes produced with this interferometer. They are obtained with monochromatic light when the adjustment of the mirrors is very exact. To understand their origin let us refer to the above arrangement which may be regarded as equivalent to interference taking place between the rays reflected from the silvered surface of M_1 and the virtual image of silvered surface of M_2 formed by the silvered surface of A so that the real mirror M_2 may be regarded as replaced by its virtual image M_2'. M_2' is then parallel to M_1. *Due to reflections in the plate A of the interferometer, the extended source may be imagined as being situated behind the observer at S and forming two virtual sources S_1 and S_2 in M_1 and M_2 respectively.*

These virtual sources originating from the division of amplitude are *coherent* as the phases of the corresponding points in the two are exactly the same at all instants. If d be the distance between M_1 and M_2', then the distance between the virtual source S_1 and S_2, *i.e.*, $S_1S_2 = 2d$. When this path difference $2d$ is exactly an integral multiple of whole wavelength, all the rays reflected normally from the mirrors will be in phase. The path difference between the two rays coming to the eye from the corresponding points S_1 and S_2 is S_1G. It is equal to $S_1S_2 \cos\theta = 2d\cos\theta$, where θ is the angle which the reflected beams make with the normal.

Since M_1 is parallel to M_2', the two rays are obviously parallel and thus angle θ is the same for the two rays. To receive the parallel rays a small telescope can be employed with advantage, especially for large values of d.

For the angle θ satisfying the relation $2\,d\cos\theta = n\lambda$, the rays would reinforce each other to produce a maximum. Thus for a given value of d, n and λ the angle θ will be constant so that the maximum lies in the form of a circle about the foot of perpendicular from the eye to the mirrors. In the line of sight of the eye perpendicular to the mirrors, the fringes are governed by the relation

$$2d = n'\lambda, \qquad \dots (i)$$

where n' is an integer. While in any other direction the relation

$$2d\cos\theta = n\lambda,$$

holds good.

From (*i*) and (*ii*), we have

$$2d - 2d\cos\theta = n'\lambda - n\lambda,$$

$$\therefore\ 2d\,(1 - \cos\theta) = (n' - n)\lambda = n''\lambda, \quad \ldots\ [n < n']$$

where n'' is also an integer.

$$\therefore \quad (1 - \cos\theta) = \frac{n''\lambda}{2d} \quad \ldots (1)$$

Now $$\cos\theta = 1 - \frac{\theta^2}{2!} + \frac{\theta^4}{4!} \ldots$$

$$\therefore 1 - \left(\frac{\theta^2}{2!} + \frac{\theta^4}{4!} \ldots\right) = \frac{n''\lambda}{2d}$$

Neglecting $\frac{\theta^4}{4!}$ and the higher terms, we have

$$\frac{\theta^2}{2!} = \frac{n''\lambda}{2d},$$

or $$\theta^2 = \frac{n''\lambda}{d},$$

or $$\theta \propto \sqrt{n''}. \quad \ldots (2)$$

If θ_1, θ_2, θ_3 be the angles for first, second, third ....rings and r_1, r_2 r_3 .... be their respective radii, then,

$$\theta_1 = \frac{r_1}{x},$$

$$\theta_2 = \frac{r_2}{x},$$

$$\theta_3 = \frac{r_3}{x}, \ldots$$

where x is the distance of M_2' from S.

Thus the radii of the rings are proportional to θ, and since by equation (2) above θ is proportional to the square root of natural numbers we conclude that the radii are proportional to the square roots of the natural numbers as in the case of Newton's rings.

Keeping M_2 fixed, if M_1 is moved backwards or forwards by half a wavelength, the change in the optical path will be λ and a circular fringe would appear or disappear at the centre of the field of view. *When, however, M_1 and M_2 are exactly coincident the central dark fringe spreads out to cover the whole field of view which will, therefore, be dark because it is clear that under these conditions the path difference will be zero for all angles of incidence.*

Fringes of the type described in this section, where parallel beams produce interference with a phase difference determined by the angle of inclination θ , are referred to as fringes of equal inclination. In contrast to the type described, this type may remain visible over very large path differences.

Example. *If the path difference between the mirrors of a Michelson's interferometer is 5 mm, what will be the angular radius of the 8th bright fringe in the circular pattern of the fringes obtained with green light of wavelength 5×10^{-5} cm.*

Using relation (1), we have

$$1 - \cos\theta = \frac{n''\lambda}{2d}$$

Here, $n'' = 8$;

$d = 5 \text{ mm} = 0.5 \text{ cm}$

and $\lambda = 5 \times 10^{-5}$ cm.

Since θ is small, putting

$$\cos\theta = 1 - \frac{\theta^2}{2!} \text{ approximately, we have}$$

$$\frac{\theta^2}{2!} = \frac{n''\lambda}{2d}$$

or $$\theta^2 = \frac{n''\lambda}{d} = \frac{8 \times 5 \times 10^{-4} \text{ cm}}{0.5 \text{ cm}} = 0.0008,$$

whence $$\theta = 0.02828 \text{ radian}$$

$$= \mathbf{16.9°.}$$

Localized Fringes

When the mirrors M_1 and M_2 in the Michelson's interferometer are not exactly at right angles to each other ($d = 0$ and $\phi \neq 1$), the mirrors M_1 and M'_2 (the image of M_2) will not be parallel, *i.e.*, the space between them forms a wedge-shaped film of air and we get localized fringes which are practically straight, parallel to the line of intersection of M_1 and M_2'. It should be noted that with a wedge-shaped film the locus of the points of equal thickness is a straight line parallel to the edge of the wedge. Since light is incident on the film at different angles, the fringes actually are slightly curved with convexity towards the edge of the wedge. As d decreases the fringe system will be displaced laterally, a new fringe crossing the centre of the field each time d is changed by $\lambda/2$. The curvature of the fringes goes on decreasing and when the path difference becomes zero, *i.e.*, we reach a point where M_1 and M'_2 actually intersect, the fringes become practically straight, that corresponding to the point O being perfectly straight and dark because the path difference for it equals $\lambda/2$ on account of reflection under different conditions and beyond this point they begin to curve in the opposite direction.

White light fringes in Michelson's interferometer. When a white light source is used in a Michelson's interferometer with the back surface of A unsilvered the central fringe, *i.e.*, the fringe corresponding to zero path difference is dark and not white as in other arrangements such as the bi-prism. This is on account of the different conditions under which reflection of the two interfering beams occurs at O resulting in phase difference of π between them.

If the back surface of A is thinly silvered the phase changes on reflection of the two beams at O are similar and the central fringe may be white. It is surrounded on either side by 8 or 10 coloured fringes since white light contains all wavelengths, between 3800 and 7500 Å provided the path difference between the two interfering beams is of the order of a few wavelengths. Beyond these few fringes *there is so much overlapping of colours* that the resultant colour is essentially white. These fringes are important because they are used to locate the position of zero path difference as required in some experiments with this interferometer. It is, however, quite difficult to locate the position of the central dark fringe by using white light only and it is best located approximately by finding the place where the localized fringes in monochromatic light become straight.

Applications of Michelson's interferometer. Michelson's interferometer has been put to many important uses. Its principal advantage lies in the fact that its two beams are widely separated and the path difference between them can be easily varied by moving the mirror M_1 or by introducing thin plates of a refracting material in the path of one of the beams. Thus it very conveniently permits accurate measurements of the distances for wavelengths and also of the refractive indices of the material of the plates put in the path of one of the interfering beams.

"Bending of a beam or even of a brick wall under the pressure of the hand can be made visible by attaching M_1 directly to the beam or wall."

We describe below a few of the important applications.

Determination of the wavelength of monochromatic light. The interferometer is properly adjusted and circular fringes are obtained with monochromatic light whose wavelength is to be determined. Telescope is focused on one of the fringes. When the mirror M_1 is moved slowly backward or forward, each fringe gets displaced parallel to itself in the field of view, the number of fringes which cross the centre of the field of view gives us a measure of the distance the mirror has moved in terms of wavelength. When the mirror moves through a distance $\lambda/2$, one fringe shifts and there is a corresponding change of λ in the path difference.

If n fringes shift across the cross-wire, when the mirror is moved through a known distance d, then the path difference is $2d$.

$$\therefore \quad 2d = n\lambda, \quad \text{or} \quad \lambda = 2d/n \qquad \text{... (3)}$$

This gives a most accurate method of determining λ as d can be measured to an accuracy of 10^{-5} cm. Further n can be increased considerably as circular fringes can be obtained for very large path differences.

Example. *In moving the mirror M_1 of Michelson's interferometer through a distance of 0.4220 mm, 1500 fringes are counted. Calculate the wavelength of light.*

When the mirror M_1 is moved backwards through a distance d, the total path difference introduced between the two beams = $2d$.

If n fringes move across the field, then

$$2d = n\lambda,$$

where λ is the wavelength of light used.

Here, d = 0.4220 mm = 0.0422 cm ;

n = 1500.

$$1500\,\lambda = 2 \times 0.0422 \text{ cm}$$

$$= 0.0844 \text{ cm.}$$

$$\text{or} \qquad \lambda = \frac{0.0844}{1500} \text{ cm}$$

$$= 5.627 \times 10^{-5} \text{ cm}$$

$$= \mathbf{5627\ Å.}$$

Determination of refractive index of a thin film. When a thin film of a transparent substance of thickness t and refractive index μ is introduced into the path of one of the interfering beams in the interferometer, the optical path in this beam is increased from t to μt. Since the beam traverses the film twice, this will introduce an extra path difference of $2(\mu - 1)t$ between the two beams, Hence the fringe system will shift. If n be the number of fringes by which the fringe system is displaced and λ the wavelength of the light used, then

$$2(\mu - 1)t = n\lambda,$$

$$\text{or} \qquad \lambda = \left[\frac{n\lambda}{2t} + 1\right] \qquad \text{... (4)}$$

Since with the monochromatic light all the fringes are exactly similar, it is not possible to distinguish one fringe from the other and as there is a sudden shift of the fringes on the insertion of the glass plate in the path of one of the beams the number n of fringes thus displaced cannot *be* counted. To overcome this difficulty white light fringes are super-imposed over the monochromatic light fringes and the central white fringe serves as the reference. The cross-wires in the telescope are focused on the central white fringe without the film and after inserting the film the number of fringes of the monochromatic light between the cross-wires and the new position of the central white fringe is counted. This gives the value of n. The accuracy of this experiment is limited by the method of measuring the thickness t of the plate. In this experiment if μ is known, t may be calculated.

For thicker plates two plates of indentical thickness are

inserted, one in the path of each beam. One of them is turned gradually about a vertical axis and the number of monochromatic fringes for a given angle of rotation are counted. From the angle of rotation the increase in thickness and hence the thickness can be calculated.

This method can also be used to study the variation of refractive index of a gas with pressure and temperature.

Example. *A thin film of glass of refractive index 1.50 is inserted in one arm of a Michelson's interferometer. Find the thickness of the film if a shift of 10 fringes is obtained, assuming the wavelength of light to be 5890 × 10^{-8} cm.*

Let t be the thickness of the film. Its insertion in the path of one of the interfering beams produces an additional path difference of $2(\mu - 1)t$, where μ is the refractive index of the material of the film. If n is the number of fringes shifted, then

$$2(\mu - 1)t = nl.$$

Here

$$\mu = 1.50;$$

$$\lambda = 5890 \times 10^{-8} \text{ cm};$$

$$n = 10$$

$$\therefore \quad 2(1.5 - 1)t = 10 \times 5890 \times 10^{-8} \text{ cm},$$

$$t = \frac{10 \times 5890 \times 10^{-8}}{2 \times 0.5} \text{ cm}$$

$$= 0.000589 \text{ cm}$$

$$= \mathbf{0.00589\ mm.}$$

Resolution of spectral lines. Michelson's interferometer has proved very useful in resolving many composite (close) lines by studying the changes in the visibility of the fringes with the increase in the path difference. Suppose a source of light S gives out a composite light consisting of two wavelengths

λ_1 and λ_2 which differ slightly ($\lambda_1 > \lambda_2$). Each wavelength gives rise to its own system of fringes which, when the two paths OM_1 and OM_2 are equal, will occupy practically identical positions. As the mirror M_1 is moved back gradually the fringe systems soon get out of step and produce positions of maximum and minimum distinctness in the field of view. At the position of maximum distinctness a bright fringe of one system falls over a bright fringe of the other system and at the position of minimum distinctness a bright fringe of one system coincides with the dark fringe of the other system. In the case of monochromatic light, however, the general appearance of the field will remain unaltered, and it remains distinct over widely different path differences. This change in the visibility of the fringes, thus, provides a means of distinguishing between light which is truly monochromatic and that which consists of two or more wavelengths so nearly equal.

Visibility of the fringes. Michelson found that clearness of fringes varies in a way characteristic of the source used. He defined the visibility of the fringes to be

$$V = \frac{I_{max} - I_{min}}{I_{max} + I_{min}} \qquad \dots (5)$$

where I_{max} is the relative intensity for a bright fringe and I_{min} is the relative intensity for the neighbouring dark fringes. Michelson devised an ingenious method* which consists essentially in plotting the variation of visibility V graphically as a function of d, the difference of optical path between the beams traversing the two arms of the interferometer. By studying the changes in the visibility of fringes he was able to resolve several composite lines. The values of V for fringes near the centre of the field obtained with cadmium red light against d. These graphs are in fact the resultant intensity graphs of a number of separate intensities which can be analysed with the help of specially designed mechanical analysers such as microphotometers, and the various components are determined.

Observations of the variation of visibility may also be used to determine the ratio of the amplitudes of the two wave systems and hence their relative intensities. *Two sine waves of equal amplitudes gave zero visibility when the waves were exactly out of step*. If the amplitudes are not equal, the visibility is never zero since the weaker fringes do not completely annul the stronger ones. If the amplitudes of the sine waves are a_1 and a_2 the maximum visibility is unity and the minimum visibility

$$V_{min} = \frac{a_1^2 - a_2^2}{a_1^2 + a_2^2} \quad \text{... (6)}$$

Thus by measuring (*i*) the separation of maxima, and (*ii*) the ratio of the maxima to the minima in the visibility curve it is possible to determine both the ratio of the wavelengths and the ratio of their amplitudes. If we apply this method of analysis, considering only the alternations, we find that yellow sodium light contains two components whose wavelengths differ by about 1 part in 1000. One contributes about twice as much as the other to the total intensity of the source. The visibility curve for the cadmium red light. Does not show any alternation of intensity but a gradual decrease of approximately exponential form. It enabled Michelson to infer that it was most nearly monochromatic of all those which he examined. The gradual decrease in the visibility curves in both cases has also been satisfactorily explained.

The arrangement also enables us to actually measure the difference between the two components λ_1 and λ_2 of a composite beam of light. To do so the position of the mirror M_1 is recorded when the centre of one maximum distinctness is on the cross-wires. On moving the mirror further backwards the distinctness decreases because the two systems start getting out of step, becomes almost zero and again improves till the centre of next maximum distinctness is brought on the cross-wires. This will occur when the optical path due to the displacement of the mirror is increased by unity in the order of the light of shorter wavelength. The position of the mirror is recorded again. If d be the distance through which the mirror is moved for the two consecutive positions of maximum distinctness (or indistinctness), the corresponding path difference is $2d$. Suppose n is the number of fringes of longer wavelength λ_1 lying in this path, then $(n + 1)$ would be the number of fringes of shorter wavelength λ_2 covering the same path. Thus

$$2d = n\lambda_2 = (n + 1)\,\lambda_2$$

or

$$n = \frac{2d}{\lambda_1}$$

and

$$n + 1 = \frac{2d}{\lambda_2}.$$

$$\therefore \qquad \frac{2d}{\lambda_2} - \frac{2d}{\lambda_1} = 1$$

or

$$2d\left[\frac{1}{\lambda_2} - \frac{1}{\lambda_1}\right] = 1,$$

whence

$$\lambda_1 - \lambda_2 = \frac{\lambda_1\lambda_2}{2d}$$

or

$$\Delta\lambda = \frac{\lambda_1\lambda_2}{2d} = \frac{\lambda^2}{2d} \qquad \text{... (7)}$$

where λ is the mean of λ_1, λ_2.

For greater accuracy a number of consecutive positions of maximum distinctness are noted and the mean value of d is determined.

This equation shows that larger the path difference between the two consecutive maxima, lesser the value of $(\lambda_1 - \lambda_2)$ and thus more monochromatic the light is. This can also be expressed as that the *larger the path difference over which the fringes remain distinct-and well-defined the more monochromatic the light is.*

***Example.** In an experiment with Michelson's interferometer the scale readings for a pair of positions of maximum indistinctness were found to be 0.6939 mm and 0.9884 mm. If the mean wavelength of the two components of the sodium D-light which was used, be 5893 A, deduce the difference between the wavelengths of the components.*

Here the displacement of the movable mirror,

$$d = (0.9884 - 0.6939) \text{ mm.}$$

$$= 0.2945 \text{ mm} = 0.02945 \text{ cm.}$$

Mean wavelength,

$$\lambda = 5893 \times 10^{-8} \text{ cm.}$$

The difference between the wavelengths λ_1 and λ_2 of the light is given by

$$\lambda_1 - \lambda_2 = \frac{\lambda^2}{2d}$$

$$= \frac{\left(5893 \times 10^{-8} \text{ cm}\right)^2}{2 \times 0.02945 \text{ cm}}$$

$$= 5896 \times 10^{-11} \text{ cm}$$

$$= 5.896 \times 10^{-8} \text{ cm}$$

$$= \mathbf{5.896\text{Å}.}$$

Measurement of the standard meter in terms of the wavelength of light. The wavelength of a spectral line is an invariable length and is very well-suited for the standardization of the meter. This was probably the most outstanding measurement made by Michelson in 1895 with the interferometer. He employed wavelengths of intense red, green and blue lines of cadmium for this purpose. Since it is physically impossible to count directly the number of fringes which shift across the field of view when the movable mirror of the inteferometer is displaced from one end of the standard meter to the other, Michelson devised a series of nine sub-standards known as etalons, each approximately twice the length of the preceding one. The length of the longest etalon was 10 cm and that of the smallest $10/2^8$ cm = 0.039 cm. An etalon consists of a bronze bar B on which two front-silvered plane mirrors M_1 and M_2 are fixed. The planes of the mirrors are kept exactly *vertical* and perfectly *parallel* to each other. The number of wavelengths contained in the distance between the two mirrors of the shortest etalon was first counted directly. Next the difference between the length of the second etalon and twice the length of the first was found. This process was repeated with successive etalons and ultimately the number of wavelengths in the 'standard meter' was determined. The procedure adopted was as follows:

The two shortest etalons (I and II) were first of all mounted side-by-side in a specially designed and modified form of the interferometer whose field of view covered all the four mirrors M_1, M_2 and M_3, M_4 of the two etalons. For this purpose an extra mirror B is employed as shown. The etalons and the mirror M could be moved parallel to each other, with their faces remaining parallel and vertical.

First of all the mirrors M, M_1 and M_2, were made parallel and coplanar with the help of white light fringes. This also made the distance of M, M_1 and M_3 from the eye equal.

White light was then replaced by the light of one of the cadmium lines, the mirror M was moved slowly from the position P to Q and the number of fringes passing the cross-wires was counted till the mirror M had reached Q when M was co-planar with M_2. This was judged by the appearance of the white light fringes in the mirror M_2, the upper mirror of the first etalon. Half the number of fringes displaced across the field of view is the number of wavelengths contained in the distance M_1M_2. The fraction of the fringe in excess, if any, was also estimated by rotation of the compensating plate C. Knowing the angle through which the compensating plate was rotated and also the amount of rotation required to produce a displacement of one fringe width, the fractional excess could be calculated to within 005 of a fringe width. Thus the distance $M_1M_2(=d)$ was determined in terms of wavelengths. The etalion 1 was then moved back through its own length till white light fringes again appeared when M_1 was at M_2 and M at Q, *i.e.*, M_1 at M_2 and M at Q were made co-planar. Finally, M was shifted to R when the white light fringes appeared in M_4 as well as M'_2 — the position occupied by M_2. The fractional excess to make M co-planar with M_4 was determined in terms of the fringes shifted. Thus the number of times the shortest etalon was contained in the next sub-standard was determined, and the exact number of wavelengths in the second etalon obtained.

The second etalon was in turn compared with the third of approximately double the length of the second and the process was repeated with successive etalons until the number of wavelengths contained in the ninth etalon, which was 10 cm in length, was determined.

Now to find the number of wavelengths of the given light contained in the *standard meter*, 10 cm etalon was compared directly with a copy of the same. This copy had two diamond scratches T_1 and T_2, distant as nearly as possible equal to one meter and was placed alongside and parallel to the 10 cm etalon. Projecting at right angles from the 10 cm etalon was an arm X which carried a similar scratch. This scratch was made to coincide with the fiducial mark T_1 on the standard meter with the aid of a powerful microscope. The movable mirror M was made co-planar with the upper mirror M_2 as observed with white light fringes. The etalon was next shifted through its own length so that in the new position of M the lower mirror M_1 of the etalon was co-planar with the above position of M. M was moved again until the white light fringes appeared in M_2, the upper mirror of the etalon. This process was repeated by alternately centring the white light fringes in the upper and lower mirrors, each time the etalon was moved through its own length. After ten such displacements of this etalon the scratch on the arm X came nearly into coincidence with the second fiducial mark T_2. The mirror M was made co-planar with the upper mirror M_2 of the etalon.

The number of fringes for bringing this scratch to exactly coincide with the fiducial mark T_2 as seen through the microscope was finally determined. In this manner the length of the standard meter was determined in terms of wavelength of the monochromatic light used. The final comparing process with the 10 cm etalon involves an accumulative error which is not much as compared with the one made in setting on the end marks T_1 and T_2 with the help of a microscope and according to the estimates of the experimenters did not exceed two fringe-widths. The final result being the mean of several sets of observations would involve a much smaller error than this. The precision of this result is better than one part in a million and it represents one of the most precise physical measurements ever attempted. The final results with the three cadmium lines were,

Red line $\quad 1\text{ m} = 1{,}553{,}163.5\ \lambda_r$ or $\lambda_r = 6438.4722$ Å.

Green line $\quad 1\text{ m} = 1{,}966{,}249.7\ \lambda_g$ or $\lambda_g = 5085.8240$ Å.

Blue line $\quad 1\text{ m} = 2{,}083{,}372.1\ \lambda_b$ or $\lambda_b = 4799.9107$ Å.

Thus the standard meter has been determined in terms of an invariable unit, the wavelength of light. Also red cadmium line has been established as the primary standard in spectroscopy.

Multiple Beam Interference

The most important application of multiple beam interferometry is the Fabry-Perot interferometer. This apparatus consists of two glass plates E_1 and E_2 separated by a distance d. Their inner surfaces are optically plane, accurately parallel and thinly silvered which reflect about 70% of incident light. The outer faces of the plates are also parallel to each other but inclined to their respective inner faces. It is to avoid interference effects due to multiple reflections and refractions that all the faces are not made parallel.

S_1 is a broad source of monochromatic light and L_1 a collimating lens which renders the incident rays parallel. Each parallel beam suffers multiple reflection in the air film between the silvered surfaces of plates E_1 and E_2 and emerges from E_2 as a parallel beam so that for interference the rays must be brought to focus at P by a convex lens L_2. The fringes are formed in its second focal plane. These are circular fringes of equal inclination. The manner how one of them is formed. If θ be the angle of incidence on the silvered surface of E_1 and d the distance between E_1 and E_2 then the path difference between successive rays = $2d \cos \theta$. The condition for the ray to produce maxima is

$$2d \cos \theta = n\lambda \qquad \ldots(8)$$

where n is an integer giving the order of interference for the particular bright fringe.

This condition is satisfied by all the points on a circle through P with centre at O_2 on the axis O_1O_2. The value of the path difference depends upon the angle θ, so that for the maxima we shall have a series of concentric rings on the screen with O_2 as centre.

In actual interferometer one plate is kept fixed while the other is capable of motion slowly towards or away from it with the help of a slow motion screw. In this manner the thickness of the air film can be changed.

Since in this case radius of a ring is a function of the wavelength so that if there are two different wavelengths in a source there are two systems of rings whose radii can be accurately determined. *Thus the interferometer can be used to compare the two wavelengths and also for the analysis of a single spectral line.* Very sharp and narrow maxima of illumination are obtained, separated by comparatively wide bands of darkness and the resolving power of the apparatus is very great.

Here the distance between the two plates is fairly large, so that the fringes are observed almost near the normal direction. These fringes of equal inclination are Haidinger fringes.

Distribution of intensity in Fabry-Perot fringes. Here we have an infinite number of beams of equal phase differences and diminishing amplitudes so that when monochromatic light is used the relatively broad $\cos^2 \theta$ type fringes associated with bi-prism, Newton's rings, Michelson's interferometer etc., are replaced by a pattern in which very *sharp* maxima of illumination comparatively wide bands of darkness. It is evident that the rate at which the intensity of the successive parallel beams 1, 2, 3, 4 etc., decreases depends upon the reflecting and the transmitting powers of the silver-coated surfaces. If a is the amplitude of the initial incident ray and R and T represent respectively the fractions of the incident light energy reflected and transmitted at each of the surfaces, then the amplitudes of the successive transmitted beams are, $aT, aRT, aR^2T, \ldots\ldots\ldots,$ $aR^{(n-1)}T$, the fractional amplitudes being $\sqrt{R}$ and $\sqrt{T}$. Neglecting small phase changes associated with metallic reflection, the common phase difference between successive beams is

$$\delta = (2d \cos \theta) \frac{2\pi}{\lambda},$$

where $2d \cos \theta$ is the path difference between them.

The disturbance at P may be represented by

$$A_1 = aT \cos \omega t,$$

$$A_2 = aTR \cos (\omega t - \delta)$$

$$A_3 = aTR^2 \cos (\omega t - 2\delta) \text{ and so on.}$$

It is possible, however, to use exponential instead of trigonometrical expressions to reduce labour. Since $e^{i\omega t} = \cos \omega t + i \sin \omega t$, the disturbance may be represented by the *real* part of $e^{i\omega t}$. Assuming δ to be the common difference of phase between the successive transmitted beams 1, 2, 3 etc. $= \frac{2\pi}{\lambda} 2d \cos\theta$; the total amplitude of the beams superimposed at P at any instant will be the sum of the real parts of the expressions,

$$aTe^{i\omega t},\ aRTe^{i(\omega t-\delta)},\ aR^2Te^{i(\omega t-2\delta)},\ \ldots \text{ etc.}$$

so that by the principle of superposition, the resultant disturbance may be represented by sum of the real part of this infinite series, that is

$$
\begin{aligned}
Z &= aTe^{i\omega t} + aRTe^{i(\omega t-\delta)} + aR^2e^{i(\omega t-2\delta)} + \ldots \\
&= aTe^{i\omega t}\left[1 + \text{Re}^{-i\delta} + R^2e^{-2i\delta} + \ldots\right] \\
&= aTe^{i\omega t} \div \left(1 - \text{Re}^{-i\delta}\right),
\end{aligned}
$$

where $\frac{aT}{1-\text{Re}^{-i\delta}}$ is the amplitude factor. But the amplitude may be the real part of either $\frac{aT}{1-\text{Re}^{-i\delta}}$ or $\frac{aT}{1-\text{Re}^{+i\delta}}$ and the intensity I may be put equal to the product of the two, so that

$$
\begin{aligned}
I &= \frac{a^2T^2}{\left(1-\text{Re}^{-i\delta}\right)\left(1-\text{Re}^{+i\delta}\right)} \\
&= \frac{a^2T^2}{1 - R\left(e^{\delta} + e^{-i\delta}\right) + R^2}
\end{aligned}
$$

$[\because e^{-i\delta} = \cos\delta - i\sin\delta \text{ and } e^{+i\delta} = \cos\delta + i\sin\delta]$

$$= \frac{a^2T^2}{1 - 2R\cos\delta + R^2}$$

$$= \frac{a^2T^2}{1+2R(1-\cos\delta)-2R+R^2}$$

$$= \frac{a^2T^2}{(1-R)^2+4R\sin^2\delta/2}$$

$$= \frac{a^2T^2}{(1-R)^2} \times \frac{1}{1+\dfrac{1}{(1-R)^2}\sin^2\delta/2} \quad \text{... (9)}$$

This expression called Airy's formula shows how the distribution of intensity in the fringe system depends upon the properties of the metal film. As $\sin^2 \delta/2$ varies between the limits 0 and + 1, so the intensity varies between a maximum value $\left(\frac{aT}{1-R}\right)^2$, when $\sin^2\frac{\delta}{2} = 0$ or $\delta = 2n\pi$

and a minimum value $\left(\frac{aT}{1-R}\right)^2 \times \frac{1}{1-\dfrac{4R}{(1-R)^2}} = \left(\frac{aT}{1-R}\right)^2$,

when $\sin^2 \delta/2 = 1$

or $\delta = (2n+1)\pi.$

Thus $I_{max} = \frac{a^2T^2}{(1-R)^2}$

and $I_{min} = \frac{a^2T^2}{(1+R)^2}$

when there is no absorption at any reflecting surface *i.e.,*

$$T = 1 - R$$

$$I_{max} = a^2 \text{ and } I_{min} = \left(\frac{1-R}{1+R}\right)^2 a^2.$$

$$\therefore \qquad \frac{I_{max}}{I_{min}} = \left(\frac{1+R}{1-R}\right).$$

We find from relation (9) that variation in intensity of the interference pattern is determined by the factor $\left[1+\frac{4R}{(1-R)^2} sin^2\ \delta/2\right]^{-1}$ in which

$$\delta = \frac{2\pi}{\lambda}(2d\cos\theta) = \frac{2\pi}{\lambda}\Delta, \text{ say.}$$

We write this factor as $\left[1+\frac{4R}{(1-R)^2}\sin^2\left(\frac{\pi\Delta}{\lambda}\right)\right]^{-1}$

If we put $R = 0.8$, the above factor becomes $\left[1+80\sin^2\frac{\pi\Delta}{\lambda}\right]^{-1}$.

When, therefore, Δ/λ is an integer, the intensity reaches a maximum value $a^2T^2/(1-R)^2$ but if Δ/λ differs from a whole number, the factor we are considering, rapidly becomes smaller than unity. Thus with a small change in Δ/λ from a whole number a great difference in intensity is produced. In other words, close on either side of a maximum, the intensity falls rapidly almost to zero and this means that maxima are sharply defined. At a distance from a maximum of one-tenth of a fringe width, the intensity falls to one sixth that of the maximum. Thus sharpness increases with reflecting power.

The expression (9) can be written as

$$I = \frac{I_{max}}{1+F\sin^2\delta/2}$$

where $$F = \frac{4R}{(1-R)^2}$$

$$\frac{I}{I_{max}} = \frac{1}{1 + F\sin^2 \delta/2}$$

The sharpness of a fringe can be expressed in terms of the width of a fringe where the intensity falls to half the maximum value. If the phase of this point differs by δ_h from that of the maximum, then

$$1 + F\sin^2 \frac{\delta_h}{2} = 2$$

or $$sin^2 \frac{\delta_h}{2} = \frac{1}{F}$$

$\therefore$ $$sin \frac{\delta_h}{2} = \frac{1}{\sqrt{F}}$$

For large values of R, δh is so small that we can take

$$sin \frac{\delta_h}{2} = \frac{\delta_h}{2}$$

$\therefore$ half width of maxima

$$\delta_h = \frac{2}{\sqrt{F}}.$$

Example. *Two Fabry-Perot interferometers have equal plate separation but values of R are 0.9 and 0.8. Deduce the relative width of the maxima in the two cases.*

Here half-width of a maxima is given by

$$\delta_h = \frac{2}{\sqrt{F}} = \frac{2(1-R)}{\sqrt{4R}} = \frac{1-R}{\sqrt{R}}.$$

For the first interferometer, $R = 0.9$

$\therefore$ $$(\delta_h)_1 = \frac{1-0.9}{\sqrt{0.9}} = \frac{0.1}{0.94} = 0.106 \text{ radian.}$$

For the second interferometer, $R = 0.8$

$\therefore$ $$(\delta_h)_2 = \frac{1-0.8}{\sqrt{0.8}} = \frac{0.2}{0.9} = 0.22 \text{ radian.}$$

∴ relative widths of the maxima

$$= \frac{(\delta_h)_1}{(\delta_h)_2} = \frac{0.106}{0.22} = \mathbf{0.5.}$$

The variation of intensity with phase (or path) difference for several values of reflection coefficient. It will at once be evident that many interfering rays give rise to *narrow maxima.*

The visibility of the fringes has already been defined as

$$V = \frac{I_{max} - I_{min}}{I_{max} + I_{min}}.$$

Hence for the Fabry-Perot interferometer the visibility is

$$V = \frac{2R}{1 + R^2} \quad \text{... (11)}$$

and thus depends only on the reflecting power of the metal surfaces and is *independent* of their transmission coefficient.

For the distribution of intensity for a finite number of beams with equal amplitudes. There while discussing the more accurate theory of the grating we have dealt with the intensity distribution for the diffraction pattern for a finite number of beams with equal amplitudes. Depicts intensity distribution for a grating with 6 slits.

Interference filters and channelled spectra. Coloured glasses or absorbing dyes have been used for many years to isolate a region of the spectrum covering a range of 500 A units or so, but in many problems in optics it is desirable to use devices to transmit narrow bands of light, a 100 A or less in width centred on a chosen wavelength within the range of visible light. Such a device is known as an *interference filter,* based as il is on the principle of interference and may conveniently be used for this purpose. In its simplest form, the arrangement very similar to a Fabry-Perot etalon, is used. Here *a* and *b* are *extremely thin,* semi transparent layers of a good reflecting material like silver separated by an optical thickness *d* equal to one or more half wavelengths of a dielectric like magnesium fluoride. To construct such a filter one layer of silver is deposited on a solid transparent substance like glass. Then a coating of the dielectric (the spacer layer) is deposited on the silver and finally the second layer of silver is added. The thin 'sandwich' of dielectric metal constitutes the filter. To protect the filter from mechanical abrasions a second glass plate is added. With such an arrangement an incident ray *PA* because of multiple internal reflections produces a bundle of emergent parallel rays. The path difference between each successive pair is $2\mu d \cos\theta$ or $2\mu d$ when light is incident normally. The emergent rays are all in phase when $2\mu d$ is equal to one wavelength λ_0 or a multiple of λ_0, *i.e.*, $2\mu d = m\lambda_0$.

With white light the intensity of the transmitted beam will be maximum for only those wavelengths which satisfy this condition. A spectrum of a parallel beam of white light which passes through the filter consists of a series of sharp bright bands separated by wide dark regions. These narrow bands are grouped around specific wavelengths. The maxima will occur at wavelengths given by the equation, $\lambda_0 = \dfrac{2\mu d}{m}$, where m is any whole number. If the thickness of the dielectric is such that

$$2\mu d = \lambda_0$$

and

$$\lambda_0 = 5000 \text{ Å}$$

then only a single narrow band in this region is transmitted because other peaks in the graph would occur at 2500 Å. 1670 Å etc., and these would be absorbed by the filter. If, however,

$$2\mu d = 4\lambda_0$$

and

$$\lambda_0 = 1250 \text{ Å}$$

then there would be transmitted beams in the visible region at 5000 Å ($4\lambda_0$) and also at 6250 Å($5\lambda_0$).

For a suitable value of *d*, there will be many narrow fringes in the visible spectrum. In order to separate them it is very necessary that the transmitted beam should be highly dispersed by an auxiliary spectroscope. These fringes are called *channeled spectrum, or Edser-Butler Bands.*

They have been used for accurate measurements of wave-lengths of absorption lines in the solar spectrum and in calibration of spectroscopes for the infra-red.

By using silver layers with reflection coefficient 94%, the peaks of the intensity curve may be made very sharp. Widths as small as 120 A may be obtained. Because of absorption by the silver layers, the intensity of the transmitted light is only from about 30 to 40% of the incident. This limitation is overcome by using in place of silver layers multiple dielectric layers. The layers are of equal optical thickness and alternate layers are of high and low refractive index so that each group acts as a single reflecting layer with reflectivity exceeding 90%. The loss of light by absorption is very small. The layers are deposited by vacuum evaporation. By a filter of this type it is possible to obtain a maximum transmission of 75% of the incident light (of that wavelength which the filter passes) and half-width about 25 Å.

Lummer-Gehrcke Plate

This is another application of multiple beam interference for producing sharp-narrow high order bright fringes. The instrument has a high resolution and is employed for observing the fine structure of spectral lines. It consists of an optically plane parallel plate *ABCD* of glass or quartz about ten to twenty cm in length, one or two cm wide and a few mm thick. The effectiveness of the instrument lies in the fact that the sides *AB* and *CD* of the plate are optically-plane and parallel. A right-angled prism *P* is cemented at one end. A parallel beam of light from a source *S* enters the prism and after reflection is incident at the inner surface *CD* of the plate at an angle slightly less than the critical angle for the material of the plate. By multiple reflections between the faces *AB* and *CD* parallel bundles of interfering rays emerge from either side of the plate at nearly grazing angles. Between any pair of rays there is the same path difference. The parallel bundles are brought to focus by the lens. *L* so that in its focal plane there are seen two sets of narrow bright fringes one on each side of the plate separated by a dark zone. Those fringes are also called Haidinger fringes. Because of great path difference and hence high order of interference, the resolving power is very large and so this instrument is much used for the analysis of a single spectral line and for the study of Zeeman effect.

Theory of the Lummer plate. If d be the thickness of the plate and μ its refractive index for wavelength λ, the optical path difference between successive emergent beams is given by

$$2\mu d \cos i = n\lambda,$$

where n is the integer giving the order of the fringe.

Since $$\sin \theta = \phi \sin i,$$

the above equation can be written as

$$2d\sqrt{\mu^2 - \sin^2\theta} = n\lambda,$$

or $$4d^2 (\mu^2 - \sin^2\theta) = n_2\lambda^2. \quad \text{... (12)}$$

This is the condition for the formation of principal maxima and is the fundamental equation for the Lummer-Gehrcke plate.

Jamin's Interferometer

This is another illustration of 'a division of amplitude' device. The interferometer was designed by Jamin in 1858, for the determination of the refractive indices of gases. The essential optical parts of Jamin's interferometer. It consists of two rectangular and equally thick plates G_1 and G_2 of about 2 cm thickness and perfectly plane parallel usually cut from the same optically worked glass. The *back* surfaces of the plates are silvered. In using the instrument the plates are arranged with their faces parallel or slightly inclined to each other. The plate G_1 is kept fixed while G_2 can be turned both about the horizontal and the vertical axes.

A parallel beam of monochromatic light from a broad source S is incident on the plate G_1 at A at 45° with the normal there. By reflection at the front and the back surfaces the beam is broken up into two parallel beams AB and CD which pass through another similar plate G_2 and combine along DT to produce interference fringes, if the plates are inclined at a small angle. Such fringes were first observed by Brewster in 1815 and are, therefore, known after him as, Brewster's fringes. The fringes are localized at infinity and can be observed by a telescope T in normal adjustment. It may be noted that here also the optical paths of the two beams from A to D will be equal if the plates are parallel.

Measurement of the refractive indices of gases. To find the refractive index of any gas at a given temperature and pressure take two similar tubes P and Q of equal length. Exhaust them completely and put them in the path of the two parallel beams. Place a sodium lamp in the first focal plane of a converging lens L and adjust the instrument to obtain fringes in the field of view of telescope T. Admit the gas slowly into the tube Q. The fringes start shifting in the field of view. Count the number of fringes which cross the field while the gas reaches the required pressure and temperature.

If l be the length of each tube and μ the refractive index of the gas, then the path difference between the two beams $= (\mu - 1)\, l$.

If m is the number of fringes shifted across the field of view and A the wavelength of monochromatic light used, then

$$(\mu - 1)l = m\lambda \qquad \text{... (13)}$$

whence μ can be calculated.

It has been experimentally found that the value of $(\mu - 1)$ at any given temperature is directly proportional to the density or pressure of the gas, so that

$$(\mu - 1)/\rho = \text{constant} \qquad \text{... (14)}$$

where ρ is the density of the gas. This relation is known as Gladstone and Dale's law.

This law may be used to find the refractive index μ_0 at N.T.P.

Since $\dfrac{\mu_0 - 1}{\rho_0} = \dfrac{\mu - 1}{\rho}$, we have

$$(\mu_0 - 1) = (\mu - 1)\frac{\rho_0}{\rho}.$$

By Charles' law,

$$\frac{p_0}{\rho_0 T_0} = \frac{p}{\rho T}$$

$$\therefore \qquad \frac{\rho_0}{\rho} = \frac{p_0 T}{p T_0},$$

whence substituting the value of $\dfrac{\rho_0}{\rho}$ in above, we have

$$(\mu_0 - 1) = (\mu - 1)\frac{p_0 T}{p T_0}$$

$$= \frac{m\lambda}{l} \cdot \frac{p_0 T}{p T_0} = \frac{m\lambda}{l} \cdot \frac{T}{273} \cdot \frac{76}{p} \qquad \textit{cf.}\text{ Eqn. (13)}$$

$$\therefore \qquad \mu_0 = 1 + \frac{m\lambda}{l} \cdot \frac{T}{273} \cdot \frac{76}{p}. \qquad \text{... (15)}$$

Jamin's compensator. To overcome the difficulty in counting the number of fringes which quickly cross the field of view a compensator *CD*. This consists of two glass plates of equal thickness fixed on a common axis and having a small constant angle between them. They can be rotated by a single knob and are so mounted in the path of the interfering beams from the two tubes that one beam passes through one plate and the other through the, other plate. Equal retardations will be produced in the path of the two beams when the rays fall on the plates perpendicularly to the plane bisecting the angle between them and therefore no displacement of the fringes takes place. In any other position the plates displace the fringes by an amount depending upon their thickness and the angle of incidence so that by rotating the axis to which they are fixed any desired displacement of the fringes can be produced or any previous displacement compensated. There is a circular scale attached to the compensator which is previously calibrated to read the optical path difference in terms of the wavelength of the monochromatic light used in the main experiment.

Using white light in the experiment the compensator is adjusted so that the central bright fringe is brought on the cross-wires of the telescope. On admitting the experimental gas in one of the tubes, the bands will shift. The compensator is then rotated to bring the central white fringe again on the cross-wire. The difference between the two readings on the calibrated circular scale gives the optical path difference $m\lambda$ introduced between the two tubes in terms of the wavelength of light used in calibration,

whence $(m - 1)\, l = m\lambda$.

Example. *One of the tubes of Jamin's interferometer contains air at room temperature and atmospheric pressure, while the removal of air from the other corresponds to a shift of 120 fringes. If the wavelength of tight be* 6×70^{-5} *cm and the tubes are 25 cm long each, find the refractive index of air.*

Here l =25 cm ; m = 120; $\lambda = 6 \times 10^{-5}$ cm.

Using the relation $(\mu - 1)l = m\lambda$, we have

$$(\mu - 1)\ 25 \text{ cm} = 120 \times 6 \times 10^{-5} \text{ cm}$$

or $$\mu - 1 = \frac{72 \times 10^{-4}}{25} = 0.000288.$$

$$\therefore \quad \mu = 1 + 0.000288$$

$$= \mathbf{1.000288.}$$

Example. *The two tubes of a Jamin's interferometer are 25 cm long each. One contains a gas at a pressure of 10 cm of mercury and the other is evacuated. If, on removing the gas a shift of 20 fringes of green light of 5760 Å is counted, calculate the refractive index of the gas at the atmospheric pressure with the help of Gladstone and Dale's law, (μ – 1) = constant x density.*

If μ is the refractive index of the gas at 10 cm of Hg pressure ; then on removing the gas a path difference of $(\mu - 1)l$ will occur, where l is the length of the tube. If m is the number of fringes moving across the field of view, then

$$(\mu - 1)\ l = m\lambda.$$

Here, l = 25 cm, m = 20; $\lambda = 5760 \times 10^{-8}$ cm

$$\therefore \quad (\mu - 1) = \frac{20 \times 5760 \times 10^{-8} \text{ cm}}{25 \text{ cm}},$$

$$= 4608 \times 10^{-8} = 0.00004608.$$

If μ' and p' are the refractive index and density of the gas at the atmospheric pressure, then by Gladstone and Dale's law,

$$\frac{\mu - 1}{\rho} = \frac{\mu' - 1}{\rho'} = \text{constant},$$

or $$\frac{\mu' - 1}{\mu - 1} = \frac{\rho'}{\rho}.$$

By Boyle's law,

$$\frac{P}{\rho} = \frac{P'}{\rho'} \quad \text{or} \quad \frac{\rho'}{\rho} = \frac{P'}{P}.$$

Here, $P' = 7.6$ cm of Hg and $P = 10$ cm of Hg.

$$\therefore \qquad \frac{\rho'}{\rho} = \frac{76}{10} = 7.6$$

whence $$\frac{\mu' - 1}{\mu - 1} = 7.6$$

or $$\mu' - 1 = 7.6\,(\mu - 1)$$

$$= 7.6 \times 0.00004608,$$

or $$\mu' = 1 + 7.6 \times 0.00004608$$

$$= \mathbf{1.0003502.}$$

Rayleigh's Refractometer

The instrument is a modification of Young's double slit device and as such unlike the instruments described, it falls in the class of *division of wavefront'* devices described in chapter 12. It is adapted for use in measuring the refractive index of air or other gases by the principle of interference.

Light from a monochromatic source S, is rendered parallel by a collimating lens L_1 and split up into two beams by a wide double slit S_1, S_2. It is allowed to pass through two similar tubes A and B placed side-by-side and having their ends closed with plane optically worked glass widows. Next after passing through two compensating plates C and D the beams are brought to focus by a lens L_2, thus producing a set of fine interference fringes superposed on a wide diffraction pattern.

Let us suppose that the two tubes contain a gas under the same conditions of pressure and temperature. Given equal lengths the optical paths are equal. Then by means of the outlet

E in the tube A the density of the gas in the tube is changed. This brings about a change in the optical path of one interfering beam and so the fringes shift in the field of view. By noting the fringe-shift for a given change in the density (or pressure) of a gas, the refractive index μ_0 of the gas at 0°C and 76 cm pressure can be calculated from the relation

$$(\mu_0 - 1) = \frac{m\lambda}{l} \times \frac{T}{273} \times \frac{76}{p_1 - p_2} \qquad \text{... (16)}$$

where m is the fringe-shift λ the wavelength of light used, l the length of the tube, T the absolute temperature, p_1 the initial pressure and p_2 the final pressure. This relation can be easily deduced from Gladstone and Dale's law.

The purpose and the mode of working of the compensator have been explained.

Stationary light waves. After the establishment of the wave theory of light in the last century, the light waves were regarded as analogous to the sound waves ; so that it was thought that it should be possible to produce stationary waves in light similar to those in sound. It was Otto-Wiener who in 1890 conclusively demonstrated the formation of stationary waves which like the stationary waves in sound are produced by the interference between the direct and the reflected waves. This is shown by the existence of *nodes and antinodes.*

Consider a monochromatic light wave travelling from left to right. When it is reflected from the surface of a denser medium, it suffers a phase change of π and moves from right to left. The equations of these two waves travelling in opposite directions can be written as

$$y_1 = a \sin 2\pi \left(\frac{t}{T} - \frac{x}{\lambda}\right) \qquad \text{... } (i)$$

$$y_2 = -a \sin 2\pi \left(\frac{t}{T} + \frac{x}{\lambda}\right) \qquad \text{... } (ii)$$

By the principle of superposition, the resultant wave is given by

$$
\begin{aligned}
y &= y_1 + y_2 \\
&= a \sin 2\pi\left(\frac{t}{T} - \frac{x}{\lambda}\right) - a \sin 2\pi\left(\frac{t}{T} + \frac{x}{\lambda}\right) \\
&= -2a \sin\left(\frac{2\pi x}{\lambda}\right) . \cos\left(\frac{2\pi t}{T}\right). \qquad \text{... } (iii)
\end{aligned}
$$

This represents the equation of a stationary wave which does not progress but only its loops keep on opening and shrinking.

Evidently, the amplitude, $2a \sin (2\pi x/\lambda)$, of the resultant wave varies with the position of the other particles. It is zero for all values of t at the points given by

$$\sin (2\pi x/\lambda) = 0,$$

or

$$2\pi x/\lambda = n\pi,$$

i.e.,

$$x = n\lambda/2 \qquad \text{... } (iv)$$

where $n = 0, 1, 2, 3$.......etc. Such points are called *nodes,* and lie in parallel planes separated by a distance $\lambda/2$. The points midway between any two consecutive nodes where the displacement is always greater as compared to elsewhere between the same two nodes are called *antinodes.* Here the amplitude has its maximum value, *i.e.,* $\pm 2a$. The antinodes are given by

$$\sin \frac{2\pi x}{\lambda} = \pm 1.$$

$$\frac{2\pi x}{\lambda} = (2n+1)\frac{\pi}{2},$$

$$x = (2n+1)\frac{\lambda}{4} \qquad \text{... } (v)$$

where $n = 0, 2, 3,$ ...... etc. These points also lie along parallel planes separated by a distance $\lambda/2$.

Thus we conclude that when a beam of monochromatic light is incident normally on a *perfectly* reflecting surface and the reflected wave suffers a phase change of π, then due to the superposition of the incident and reflected waves stationary waves are formed in front of the reflector. The reflecting surface acts as a nodal plane and other nodal planes will be parallel planes at distances $\lambda/2$, λ, $3\lambda/2$, etc. from it. At points midway between any two consecutive nodal planes there are antinodal planes which will be parallel planes at distances $\lambda/4$, $3\lambda/4$, $5\lambda/4$, etc., from the reflecting surface. In nodal planes ether is permanently at rest while in antinodal planes it alternately suffers maximum displacement.

Wiener's Experiment

To demonstrate this phenomenon, Wiener employed the arrangement. He used a glass plate coated with a thin photographic film of a specially prepared emulsion. He placed the film side against a perfectly reflecting surface and inclined to it at a small angle so that it exactly crosses the nodal and antinodal planes successively in a stationary wave pattern formed by the superposition of direct and reflected wavetrains when a beam of monochromatic light falls upon the mirror through the film. At the nodal planes the silver halide is not affected since the ether is permanently at rest, while reverse is the case at the antinodal layers where the film is reduced to silver.

As expected when the plate was developed and examined under a high power microscope, the film was found to be crossed by a series of thin equidistant parallel dark bands of silver (antinodal layers) lying along the lines of intersection of the film with the antinodal planes; while along the lines of intersection of the film with the nodal planes there were transparent parallel layers of gelatine (nodal layer). A remarkable thing to be noted was that no blackening was found along line of contact of the film with the reflecting surface; while the first black band which corresponds to an antinodal layer was found at a distance of $\lambda/4$ from the reflecting surface. Thus the mirror forms a nodal layer. The distance between any two consecutive layers of silver is $\lambda/2$.

Colour photography. It is the photography of objects in their natural colours and is a very important application of the phenomenon of light.

Soon after the publication of Wiener's experiment in 1890, Lippmann* in 1891 developed a method of photographing objects in their true colours. This method is based upon the phenomenon of the formation of stationary waves in light discussed above.

He placed in a camera a specially prepared photographic plate backed by a suitable reflector—a layer of mercury with the glass side facing the camera objective. This was achieved by using a special form of plate holder. When the plate is exposed to a coloured object illuminated by white light, the light reflected from the object enters the film *normally*, gets reflected from the mercury surface and comes back to glass through the film ; thus setting up stationary wave pattern in the film emulsion. Each colour in the object produces its *own set* of nodal and antinodal planes. When the plate is developed and fixed, the emulsion film contains different sets of antinodal reflecting layers of metallic silver parallel to the reflecting surface, each set of antinodal layers, corresponding to a particular colour. It is because of the variation in the wavelength of incident light that these layers are not uniformly separated throughout the whole length of the plate. But, however, the separation between any two consecutive layers is $\lambda/2$, where λ is the wavelength of the incident light at the point. Thus according to different wavelengths of light different sets of reflecting planes (metallic, silver layers) located at the antinodes are formed in the emulsion.

If now the plate so photographed is illuminated with white light and is seen by reflection, the light is partially reflected at each layer and the path difference between the waves reflected from the two successive layers of one set will be one wavelength for that colour (wavelength) for which it was developed. For one set of antinodal reflection planes only those reflected waves reinforce one another's effect for which half wavelength equals the distance between two consecutive reflecting layers. So to say the path difference between the successive reflected waves of this colour equals twice the separation between the successive layers *(i.e.,* one wavelength). Thus the waves satisfy the conditions for constructive interference and reinforce one another.

In this way for different sets of reflecting planes different wavelengths satisfy the conditions of constructive interference.

Hence the layers of one set reflect strongly only that colour (wavelength) by which they were originally formed, while the other colours for which the path difference between the reflected beams is not exactly one wavelength will also be reflected but with much less intensity.

Thus when this exposed photographic plate is illuminated with white light and is seen by reflection, every set of reflecting planes reflects back strongly only that colour by which it was originally formed when it was exposed.

This method gives extremely brilliant colour photographs but the process is very difficult to carry out and more so because no colour prints can be made from the plates, as they themselves serve as positive, when viewed by reflected light. Thus this process is of very little practical use. It only serves to demonstrate the existence of stationary waves in light and thus provides another evidence of the wave character of light. From the fact that the reflecting surface serves as a node, Wiener also concluded that *it is the electric vector and not the magnetic vector which is effective in all optical phenomenon.*

Commercial processes. The modern colour photography is based upon *three-colour process.* Since all the colours in nature are obtained by a mixture of three primary colours selected from the spectrum, *red, green and blue-violet* in proper proportions, so that all the processes of colour photography seek first to analyse the different amounts of three primary colours in the object to be photographed and then combine them to produce the same sensation on the brain as the original object. To achieve this purpose three-colour filters red, green and blue are employed.

The two main processes of colour photography in use at present are :

1. the additive process, and
2. the subtractive process

but the discussion of these topics is beyond the scope of the book.

5

Division of Amplitude

The colours of thin films. Of the two classes- of devices for producing interference phenomenon, *the division of* wavefront and *the division of amplitude,* we discussed in the preceding chapter the devices belonging to the former class namely, Fresnel's bi-prism. Fresnel's double minors, Lloyd's mirror, etc. In this chapter we shall discuss a few devices which belong to the latter class of interference phenomenon. The simplest example of study of this class is the 'colours of thin films'.

It is a matter of common experience that when a thin film of oil spread on the surface of water is exposed to an extended source of white light brilliant colours are seen. Similar colour phenomenon is also exhibited by soap bubbles, and metallic surfaces when heated. These beautiful colour effects arise from interference between the light waves reflected from the two surfaces of thin films of transparent materials.

Let us explain these effects by considering an ideal case of reflection from a thin film with perfectly plane sides which are parallel to each other.

Interference Phenomena

Interference phenomenon with a plane-parallel thin film. In order to understand clearly how these interference effects are produced let us consider a thin film of thickness t and refractive index μ bounded by XY and X_1Y_1 two parallel plane surfaces surrounded by air. Suppose a plane wave of monochromatic light of wavelength λ and amplitude a from an extended source of light S, such as a sodium flame, is incident on the surface XY of the film at A, where the light is partly reflected along AP and partly refracted along AE. Upon arrival at E a part of the latter is again reflected along EB *inside* the film and a part refracted *out* of the medium towards P_1 so that EP_1 is parallel to SA. At *B the ray EB is again divided.* Similar reflections and refractions occur at F, C, G and so on. A continuation of this process yields two systems of parallel rays, one on each side of the film. In each system intensity falls rapidly from one ray to the next. If these parallel reflected or refracted rays are collected and focused by a lens, like that of the eye, interference patterns are observed on the retina, because each ray has travelled a different distance and the path difference between the two successive rays may be such as to produce constructive or destructive interference according as their phases differ by even or odd multiple of π. We shall now consider the reflected and refracted systems of rays respectively.

Reflected system. To find the path difference between a pair of successive rays AP and BQ, draw BN and AM perpendiculars on AP and EB respectively. Also draw AL perpendicular to the faces of the film and produce it to meet BE produced in G. Then, obviously, $\angle ABN = I$ and $\angle BAM = l$ $\angle EAL = \angle EGL = r$, where i and r are angles of incidence and refraction respectively. Also Δs ALE and GLE are congruent.

$$\therefore \qquad EA = EG$$

$$\text{and} \qquad AL = LG.$$

Here the optical paths from N and B to the focus of the lens will be equal. Thus if μ be the refractive index of the material of the film the path difference measured in air between the two rays AP and $AEBQ$

$$= \mu\,(AE+EB)-AN$$

$$= \mu\,(GE+EB)-AN$$

$$= \mu\,GB-AN$$

Now $$\mu = \frac{\sin i}{\sin i}$$

$$= \frac{AN/AB}{MB/AB}$$

$$= \frac{AB}{MB}$$

or $$AN = \mu\,MB.$$

$$\therefore \text{path difference} = \mu\,\text{GB} - \mu\,MB = \mu GM$$

$$= AG\cos r = \mu\,2AL\cos r$$

$$= 2\,\mu t\cos r \qquad ...(1)$$

Since the ray AP suffers reflection at an air-medium interface, it undergoes a phase change of $\lambda/2$, or a path increase of $\lambda/2$, where λ is the wavelength of light in air while the ray $AEBQ$ does not undergo any phase change, since it is reflected at E, the surface of a denser-to-rarer medium.

Thus the effective path difference between the two rays AP and $AEBQ = 2\mu t\cos r - \lambda/2$.

Now let us examine the phases of the remaining rays. From the geometry, it is evident that the total path difference between BQ and $CR = 2\mu t\cos r$, since there are only internal reflections involved. If $2\mu\text{t}\cos r = n\lambda$, the ray CR will be in phase with ray BQ. The same holds true for all succeeding pairs. Thus we conclude that for this condition to hold first and second rays will be out of phase with each other but second, third, fourth, etc., will all be in phase with one another, but their amplitudes fall off rapidly.

(*i*) If the thickness of the film is very small as compared to the wavelength of light used, so that $2\mu t \cos r$ can be neglected, then the total path difference between rays 1 and 2 will reduce to ft/2. Thus two rays *AP* and *BQ* will interfere destructively and darkness will result.

(*ii*) If the thickness of the film is not negligible,

(*a*) the film appears bright, if

$$2\mu t \cos r - \lambda/2 = n\lambda,$$

where n is a positive integer including zero.

$$\therefore \quad 2\mu t \cos r = (2n + 1)\,\lambda/2 \text{ for a maximum....(2)}$$

Here second ray will be in phase with the first, but third, fifth and seventh, etc., will be out of phase with second, fourth, sixth, etc. Since the second ray is more intense than the third ray, and fourth more intense than fifth and so on, these pairs cannot cancel each other. The rays of stronger series combine with the first, which is the strongest of all, thus giving rise to a maximum of intensity.

(*b*) the film appears dark, if

$$2\mu t \cos r - \lambda/2 = (2n - 1)\,\lambda/2,$$

or $\quad 2\mu t \cos r = n\lambda$ is the condition for a minimum...(3)

Thus a number of bright bands separated by dark intervals will be seen.

Here second ray is out of phase with the first, but first is much stronger than the second so that the two will not completely annul each other. The addition of the third, fourth, fifth, etc., which are all in phase with the second will be sufficient to produce complete darkness at the minima. Since the intensities of the two interfering beams are never quite equal and differ slightly, the fringes are never quite so pronounced.

Transmitted system. The rays emerging from the lower side of the film may also be brought together by a lens and caused to interfere.

Now proceeding in the same manner as in the reflected system and constructing the diagram as 12-3 we find that the path difference measured in air between the two transmitted rays EP_1 and $EBFQ_1$ is given by

$$= \mu\,(EB + BF) - EN$$

$$= \mu\,(EB+BG) - EN$$

$$= \mu\,EG - EN.$$

Since $\mu = \dfrac{\sin i}{\sin r}$

$$= \frac{EN/EF}{EM/EF}$$

$$= \frac{EN}{EM}$$

$\therefore$ $EN = \mu\,EM.$

$\therefore$ path difference $= \mu EG - \mu EM = \mu MG$

$$= \mu\,FG \cos r = 2\,\mu\,FL \cos r$$

$$= 2\mu t \cos r. \quad \text{...(4)}$$

Since here reflection always takes place on the surface of a rarer medium so that there is no extra path difference introduced on this account.

If the thickness of the film is very small as compared to the wavelength of light used, so that $2\mu t \cos r$ is negligible, the two waves will have zero path difference and will reinforce each other ; so that the film appears bright by transmitted light.

With the increase in the thickness of the film the path difference also increases. When it is $\lambda/2$ the film appears dark in the transmitted system (while in the reflected system it appears bright). If the path difference increases further till it becomes λ it appears bright in the transmitted system (whereas in the reflected system it appears dark). Thus a film gradually increasing in thickness appears bright if

$$2\mu t \cos r = n\lambda, \quad \text{...(5)}$$

and it appears dark if

$$2\mu t \cos r = (2n + 1)\lambda/2 \quad \text{...(6)}$$

Therefore in this case also, a number of bright bands separated by dark intervals will be seen.

It is evident from the relation deduced above for the brightness and darkness of the film in the reflected and transmitted systems that the conditions of interference for the two systems are quite opposite to each other. A bright band in the reflected system will be a dark band in the transmitted system for the same path difference and *vice-vera*. Thus the *two systems are* complementary. It may, however, be noted that in the reflected system at almost normal incidence on account of a small difference in the intensities of the two interfering beams, the intensity never vanishes at a minimum completely and the fringes are never quite so pronounced as in the bi-prism experiment. On the other hand in the transmitted system, the two interfering beams vary considerably in intensity and at normal incidence there is hardly a difference of 15 to 20% in the maximum and minimum intensities.

Colours of thin films by white light. When white light is used the film appears to be coloured as explained below :

The path difference $2\mu t \cos r$ between the interfering beams depends upon μ, t and r, so that even if t and r are kept constant, the path difference would change with μ or the wavelength of light used. If SA be a ray of violet light the thickness of the film in order that a retardation of $\lambda/2$ may occur along the path AEB, will be less than in the case of red light since $\lambda_v > \lambda_r$ and hence the distance between two bright bands will be less for violet than for red light. Since white light is composed of different colours the path difference between rays coming from any point on the film will be different for different colours, so that with white light the film shows various colours from *violet* to *red*. If the thickness of the film varies uniformly the same order of colours will be observed beginning with the thin end of the film which will appear black. *These, colours are determined by both the thickness of the film and the angle of incidence within the film.*

As the path difference varies with the thickness of the film so that it passes through various colours for the same angle of incidence when seen in white light.

Further the path difference changes with the angle r and angle r changes with angle i so that with white light the film assumes various colours when viewed from different directions.

Passing further in the direction in which the thickness of the film increases we shall ultimately reach a point where the overlapping produces uniform illumination. Further, with white light as the source, the colour of the film, either by transmitted or reflected light will be due to the suppression of certain wavelengths by interference. Since in any case the wavelengths which are suppressed in the reflected ray are present in the transmitted ray, it follows that the colour of a film when viewed by reflected and transmitted light respectively will be complementary.

Demonstrating colours of thin films. The colours of thin films can be conveniently demonstrated to a class by forming a soap film in a ring or the mouth of a glass funnel about 10 cm in diameter held in a stand horizontally. By means of muslin cloth pad dipped in the soap solution containing a small quantity of glycerine and rubbed against the mouth of the funnel it is easy to obtain a thin film stretched across it The film soon becomes wedge-shaped, its thickness gradually increasing from top downwards. It may be illuminated with light from an arc lamp and seen by the reflected light No colours are seen in the beginning when the film is thick. As the soap solution drains down brilliant colours make their appearance in the form of horizontal bands crossing the film. The top edge of the film soon becomes very thin and a black band makes its appearance. The colours below undergo a change as the film is thinning out due to drainage of solution.

The coloured bands formed by a thin soap film when viewed by reflected light were classified by Newton as follows, beginning with the black film of infinitesimal thickness.

1. Black, blue, white, yellow, red.
2. Violet, blue, green, yellow, red.
3. Purple, blue, green, yellow, red, so on.

Necessity of a broad source for observing colours of thin films. It may be observed that a broad source of light is necessary for seeing colours of thin films. To explain this point let us suppose that the film is viewed by reflected light from a point source S. Here we find that pairs of light rays 1, 2 and 3, 4 etc. reflected from different parts of the film ff are all being sent in different directions according to the laws of reflection. If the eye is kept in a fixed position E then on account of the limiting size of its pupil light from a very small portion of the film where rays 1 and 2 are reflected, enters the eye but for observing different colours it is necessary that a large portion of the film must be seen at the same time. With a broad source of light S_1 S_2 S_3 pairs of interfering rays reflected from a very large portion of the film enter the eye simultaneously, hence the colours are easily seen.

Example. *A parallel beam of sodium light* ($\lambda = 5890 \times 10^{-8}$ *cm) is incident on a thin glass plate* ($\mu = 1.5$) *such that the angle of refraction into the plate is 60°. Calculate the smallest thickness of the plate which will make it appear dark by reflection.*

Let the smallest thickness of the plate be t cm for which it appears dark when seen by reflection.

For a dark band in the reflected system

$$2\,\mu\cos r = n\lambda.$$

Since for the least thickness of the plate n must be one and hence the path difference

$$2\mu\cos r = \lambda$$

or $$t = \frac{\lambda}{2\mu\cos r}.$$

Here $$\mu = 1.5\,;$$
$$r = 60°\,;$$
$$\lambda = 5890 \times 10^{-8}\text{ cm.}$$

$$t = \frac{\lambda}{2\mu\cos r} = \frac{5890 \times 10^{-8}}{2 \times 1.5\cos 60°}\text{cm}$$

$$= \frac{5890 \times 10^{-8}}{2 \times 1.5 \times 0.5}\text{ cm}$$

$$= 3927 \times 10^{-8}\text{ cm.}$$

Example. *A soap film of* μ *=1 333, is illuminated by white light incident at an angle of 45". The light refracted by it is examined by a spectroscope and two bright bands are focused corresponding to wavelengths* 61×10^{-4} *cm and* $6\,0 \times 10^{-5}$ *cm. Find the thickness of the film.*

For consecutive bright bands in the refracted system corresponding to wavelengths λ_1 and λ_2,

$$2\,\mu\,t\cos r = n\lambda_1(n+1)\,\lambda_2$$

where t is the thickness of the film at the point and r is the angle at which light is refracted into the film.

Solving for n, we have

$$n = \lambda_{2/}(\lambda_1 - \lambda_2).$$

Hence $2 \mu t \cos r = \lambda_1\lambda_2 (\lambda_1 - \lambda_2)$,

or
$$t = \frac{\lambda_1 \lambda_2 / (\lambda_1 - \lambda_2)}{2 \mu \cos r}$$

$$= \frac{\lambda_1 \lambda_2}{2 \mu \cos r (\lambda_1 - \lambda_2)}.$$

Here, $i = 45°$, $\mu = 1.333$, $\lambda_1 = 6.1 \times 10^{-5}$ cm

and $\lambda_2 = 6.0 \times 10^{-5}$ cm.

Now, $$\mu = \frac{\sin i}{\sin r} = \frac{\sin 45°}{\sin r},$$

or $$\sin r = \frac{\sin 45°}{\mu} = \frac{0.7071}{1.333}.$$

$\therefore$ $$\cos r = \sqrt{1 - \sin^2 r} = \sqrt{1 - \left(\frac{0.7071}{1.333}\right)^2} = 0.8476.$$

Substituting the values in the above equation, we get

$$t = \frac{6.1 \times 10^{-5} \text{ cm} \times 6.0 \times 10^{-5} \text{ cm}}{2 \times 1.333 \times 0.8476 \times 0.1 \times 10^{-5} \text{ cm}}$$
$$= \mathbf{0.00162 \text{ cm}}.$$

Why interference fringes cannot be obtained with thick films ? Consider a thick film *f f*, illuminated by an extended source δ. Let E be the position of the eye focused on some point A of the film. On account of the finite size of the pupil of the eye (about 3 mm), the rays from the region *ab* of the source after incidence at A will reach the eye simultaneously. Since the rays from a and b are differently inclined, the path difference of the pair of incident rays from a is somewhat different from that of the other pair from *b*. To find its value we differentiate , $x = 2\mu t \cos r$, whence we get, $dx = -2\mu t \sin r\, dr$. Here *t* is thickness of the film and μ its refractive index.

Now, if *t* is not more than a few wavelengths the difference *dx* is a small fraction of a wavelength. If, therefore, *A* is bright as observed with one pair, it will be bright also for the other pair. If *t* is large, a few mm or more, the difference *dx* becomes very large until when it is equal to half a wavelength, one pair may give rise to a maximum and the other to a minimum or *vice versa*. In such a case interference fringes cannot be seen. For if the two waves resulting from the ray emanating from *a* annul each other at *A*, the pair from *b* will reinforce and so the effect 6*f* the beam of rays entering the eye would also be the same, that is, interference effects would cease.

Different Fringes

Fringes produced by wedge-shaped films. Let us consider a wedge consisting of two thin glass plates sightly inclined to each other at an angle θ and enclosing a liquid of refractive index μ. Let the film be illuminated by monochromatic light from a slit held parallel to the edge of the wedge. Interference occurs between the rays reflected at the upper and lower surfaces of the film respectively and '*parallel* interference bands result which are alternately bright and dark. If *t be* the thickness of the film at a distance *x* from the edge, for light incident normally the path difference is $2\mu t = 2\mu x\theta$ (if θ is very small) The condition for brightness is

$$2\mu t = \left(n + \frac{1}{2}\right)\lambda$$

or

$$2\mu x\theta = \left(n + \frac{1}{2}\right)\lambda,$$

and the condition for darkness is

$$2\mu t = n\lambda \text{ or } 2\mu x\theta = n\lambda.$$

If x_1 is the distance of the nth dark band from the edge and x_2 that of the $(n + m)$ th dark band, then

$$x_1 = \frac{n\lambda}{2\mu\theta}$$

and $$x_2 = \frac{(n+m)\lambda}{2\mu\theta},$$

or $$x_2 - x_1 = \frac{(n+m)\lambda}{2\mu\theta} - \frac{n\lambda}{2\mu\theta} = \frac{m\lambda}{2\mu\theta}$$

The width of a single band

$$\beta = \frac{x_2 - x_1}{m} = \frac{\lambda}{2\mu\theta}. \qquad \text{...(7)}$$

The same relation is obtained if we consider bright fringes.

This result affords us a method of finding (*i*) λ if μ and θ and known, (*ii*) μ when θ and λ are given, and (*iii*) θ when μ and λ are given.

If white light is used each of its constituent colours gives rise to its own fringe system but of a different width. The thin edge of the wedge appears dark for reasons already explained. Following the dark fringe we come across the reinforcement of colours from violet to red in a spectral sequence and then coloured bands with different shades formed by the superposition or interference of fringe systems. Finally, the overlapping results in uniform illumination.

It may be noted that in a wedge-shaped film for normal incidence change from maximum to minimum intensity in the interference pattern is mainly due to the change in the thickness t or the distance from the edge and *each fringe is the locus of points where the thickness of the film has a constant value.* For this reason such fringes are often known as fringes of equal thickness and are easily viewed with a magnifying glass or a low power microscope unless the wedge angle is very small.

Example. *A wedge-shaped air film having an angle of 40 seconds is illuminated by monochromatic light and fringes are observed vertically through a microscope. The distance measured between consecutive bright fringes is 0.12 cm. Calculate the wavelength of light used.*

For *nth* bright fringe, $2\mu t = \left(n + \frac{1}{2}\right)^{\lambda}$

Since it is air film, $\mu = 1$,

also $t = \theta$

$$\therefore \qquad 2\times\theta = \left(n + \frac{1}{2}\right)\lambda. \qquad ...(i)$$

For the $(n + 1)$th bright fringe, $2\delta\theta' = \left(n + 1 + \frac{1}{2}\right)\lambda$ .... *(ii)*

Subtracting *(i)* from *(ii)*, we have

$$2\theta(x' - x) = 1$$

Here, $\quad x' - x = 0.12$ cm

and $\quad \theta = 40 \text{ seconds} = \frac{40}{60 \times 60} \times \frac{\pi}{180}$ radian

$$= \frac{\pi}{90 \times 180} \text{ radian.}$$

$$\therefore \qquad \lambda = 2 \times \frac{\pi}{90 \times 180} \times 0.12 \text{ cm.}$$

$$= 4.657 \times 10^{-5} \text{ cm} = 4657 \text{ Å.}$$

Example. *Two pieces of plane glass are placed together with a piece of paper between the two at one edge, forming a wedge-shaped air film. If on viewing the film with a light of wavelength 6000×10^{-8} cm there are 10 fringes per cm, calculate the angle of the wedge. Assume normal incidence.*

If r is the thickness of the air film at a distance x from the edge, the path difference for normal incidence is $2t$ and the condition for darkness is $2t = n\lambda$.

If θ is the angle of the wedge, then $t = x\theta$, since θ is small,

$\therefore$ for a dark band, $2x\theta = n\lambda$

andfor a bright band, $2\, x\theta = \left(n + \frac{1}{2}\right)\lambda$,

$\therefore$ distance of nth dark band from the edge,

$$x_n = \frac{n\lambda}{2\theta},$$

and that of the $(n + m)$th dark band,

$$x_{n+m} = \frac{(n+m)\lambda}{2\theta}.$$

$$\therefore \qquad x_{n+m} - x_n = \frac{m\lambda}{2\theta},$$

or
$$\theta = \frac{m\lambda}{2(x_{n+m} - x_n)}$$

Here, $x_{n+m} - x_n = 1$ cm ; $m = 10$; $\lambda = 6000 \times 10^{-8}$ cm.

$$\therefore \qquad \theta = \frac{10 \times 6000 \times 10^{-8}\ \text{cm}}{2 \times 1\ \text{cm}} = 3 \times 10^{-4}\ \text{radian.}$$

$$= 3 \times 10^{-4} \times \frac{180}{\pi} \times 60 \times 60\ \text{sec.}$$

$$= \textbf{61.9 seconds.}$$

Fringes of equal inclination. If the faces of a thin film are plane parallel, the path difference $2\mu t \cos r$ between the two parallel rays is due to the variation in the angle r only. The fringes in this case are called fringes of equal inclination (Haidinger's fringes), each fringe being the locus of the points where the angle r has a constant value. These fringes will, therefore, have the form of an arc of a circle whose centre is at the foot of the perpendicular drawn from the eye to the plane of the film. Such fringes are seen with Michelson's interferometer (Art. 13-3) when the image of M_2 is set accurately parallel to M_1. They may also be seen with a glass plate several millimeters thick and of moderately good quality using normal incidence. An extended source of light is *essential* for their production to get a range of r values each being due to a parallel beam of light which has its origin in a different point of the source. To observe these fringes the eye must be focused for parallel light or a telescope in normal adjustment may be used, whereas fringes of equal thickness described above can be seen by the eye (or microscope) focused on the film.

The transmitted system gives us similar results.

Testing a surface for plane lens. The interference fringes obtained from a film of varying thickness have been usefully applied in the testing of optically plane surfaces. If wedge-shaped air film is formed between an optically plane glass plate and the surface under test the fringes will be straight if the surface is perfectly plane, otherwise irregular in shape. From the formula $2\mu t = \left(n + \frac{1}{2}\right)\lambda$, we find that a fringe is characterized by a particular value of n and hence will follow those parts of the film where t is constant. The contour of a particular fringe would indicate the same thickness of the air film.

Newton's Rings

This phenomenon is a special case of interference in a thin film of slowly varying thickness and is frequently employed in the laboratory for the measurement of wavelength of monochromatic light and refractive indices of rare liquids. A mention of this phenomenon was first made by Hooke in 1665.

Newton studied the subject more fully by making careful measurement of diameters of the circular rings produced in the air film enclosed between the slightly curved surface of a convex lens and a plane glass plate in contact with it and hence they are known as Newton's rings. These results were interpreted later by Young as being due to interference.

Newton's rings by reflected light. An experimental arrangement for producing and observing Newton's rings by reflected light. Light from a monochromatic source S rendered parallel by a convex lens L_1 and reflected by a glass plate G held at 45° to the rays, is made to fall normally on a long focus plano-convex lens L_2, which encloses a thin film of air or a rare transparent liquid between its convex side and a plane glass plate P. Light transmitted through the lens L_2 on reflection from the surface of the glass plate P interferes with the light reflected from the lower surface of the lens. These beams of light after reflection proceed upwards and enter the observer's eye through a low-power travelling microscope M. On focusing the microscope at the point of contact of the lens and the plate a large number of rings, alternately dark and bright, is seen. The central ring is dark. The outer rings gradually decrease in width with the increase in radius. Since the air film is symmetrical about the point of contact C, the rings will be concentric with centre at C *because they follow lines of equal thickness.*

Theory. To obtain an expression connecting the diameter of a ring with the radius of curvature of the convex surface, the wavelength of light and the order of the ring, let us consider the vertical section $ACBE$ of the plano-convex lens through its centre of curvature. Suppose R is the radius of curvature of the convex surface of the lens and C the point of contact of the lens with the plane surface such that the points A and B are equidistant from C.

Draw AA_1 and BB_1 perpendiculars to the plane A_1B_1 of the plate P. Let $AA_1 - BB_1 = t$ be the thickness of the air film at a distance r (= $CA_1 = CB_1$) from the point of contact C of the lens with the plate.

From the geometry of the circle, we have

$$DE \times CD = AD \times DB$$

$$\text{or} \quad (CE - CD)CD = AD^2, \qquad [\because AD = DB]$$

$$\text{or} \quad (2R - t)\, t = r^2,$$

where r is the radius of the ring.

Since the radius of curvature R of the lens is very large, and t is very small, we can neglect t as compared to $2R$.

$$\therefore \quad 2Rt = r^2 = \left(\frac{AB}{2}\right)^2 = \left(\frac{d}{2}\right)^2 = \frac{d^2}{4}$$

$$\therefore \quad 2t = \frac{d^2}{4R} \qquad \ldots (8)$$

where d is the diameter of the ring.

The path difference between the two rays, one reflected from A and other from A_1

$$= 2\mu t \cos \phi \qquad \ldots (9)$$

where ϕ is the angle of refraction. It is $2t \cos \phi$ if it is an air film.

Since the ray reflected at A_1 or B_1, the air medium surface suffers a phase change of π and hence a further increase in path by $\lambda/2$ the effective path difference between the interfering beams becomes $2t \cos \phi + \lambda/2$.

Thus the points A and B will lie on a bright ring of diameter AB if the path difference

$$2t \cos \phi + \frac{\lambda}{2} = n\lambda,$$

or $$2t \cos \phi = (2n-1)\frac{\lambda}{2}.$$

Since the rays are incident practically normally, the angle ϕ is nearly zero, and hence $\cos \phi = 1$.

$$2t = (2n-1)\frac{\lambda}{2} \quad \text{... (10)}$$

Substituting the value of $2t$ from eqn. (8) in the above equation, we get for the nth bright ring,

$$\frac{d_n^2}{2R} = (2n-1)\frac{\lambda}{2},$$

or $$d_n^2 = (2n-1)\ 2R\lambda,$$

or $$d_n = \sqrt{(2n-1)2R\lambda}. \quad \text{... (11)}$$

As R and λ are constants, so that $d_n \propto = \sqrt{(2n-1)}$ where n has values 1, 2, 3... for first, second, third ... rings respectively. Thus the diameters of the bright rings are proportional to the square roots of odd natural numbers.

Also for the nth dark ring,

$$\frac{d_n^2}{2R} = n\lambda,$$

or $$d_n = \sqrt{4Rn\lambda}, \quad \text{... (12)}$$

$$\therefore \quad d_n \propto \sqrt{n}, \quad \text{where } n = 1, 2, 3 \text{ etc.}$$

Thus the diameters of the dark rings are proportional to the square roots of the natural numbers.

We shall consider at this stage a point of interest about interference in thin films. You have noticed that Newton's rings are only seen by focusing the microscope on the thin film which is responsible for their production. They are, therefore, known as localized fringes and are distinguished from the interference fringes obtained with the aid of Young's slits, Fresnel's mirrors or Lloyd's mirror, *i.e.*, those produced by two coherent sources, by the fact, that the latter type of fringes can only be observed at places where two trains of waves actually cross. But there is a wide region of space where such crossing occurs, *i.e.*, the fringes extend into space so that it is not necessary to put the observing screen or focus the microscope on any special place. For that reason this latter class of fringes is called non-localized fringes.

Newton's rings between two curved surfaces. If the air film is formed between two curved surfaces, the formula for the nth bright ring will be modified as under :

Let R_1 and R_2 ($R_2 > R_1$) be the radii of curvature of the two surfaces in contact at O and r_n the radius of the nth ring, then for normal incidence,

$$P_nB_n = \frac{r_n^2}{2R_1}$$

and $$Q_nB_n = \frac{r_n^2}{2R_2} \quad \textit{cf.}\text{ equation (8) above}$$

$\therefore$ the apparent path difference $(P_nB_n - Q_nB_n)$

$$= \frac{1}{2} r_n^2 \left(\frac{1}{R_1} - \frac{1}{R_2}\right)$$

For the ring to be bright, $2(P_nB_n - Q_nB_n) = (2n - 1)\dfrac{\lambda}{2}$

$$r_n^2 \left(\frac{1}{R_1} - \frac{1}{R_2}\right) = (2n-1)\frac{\lambda}{2} \qquad \text{... (13)}$$

If the surface of radius of curvature R_2 be convex as seen from above, then

$$r_n^2\left(\frac{1}{R_1}-\frac{1}{R_2}\right) = (2n-1)\,\frac{\lambda}{2} \qquad \ldots (14)$$

If white light is used then because of overlapping of the rings of all the different wavelengths in white light, only a few rings and those too highly coloured, except the central spot which is black, are observed. It is because the diameters of the rings depend upon the wavelength of light used.

Newton's rings by transmitted light. Newton's rings can also be obtained by the transmitted light. The interference fringes in the transmitted system are produced by the interference of the two transmitted rays BC and B_1C_1. The ray BC originates from AB by single refraction at B while the ray B_1C_1 also originates from AB by two reflections, one at B and the other at A_1, and then a refraction at B_1. At each reflection a phase change of π is introduced so that a total phase change of $\pi + \pi = 2\pi$ is introduced in B_1C_1, which physically means no phase change. Thus the only phase difference between the two refracted rays BC and B_1C_1 is due to the difference in their optical paths traversed.

If the angle of refraction of the transmitted rays be ϕ, then the effective path difference is $2t \cos \phi$.

$\therefore$ for bright rings,

$$2t \cos \phi = n\lambda, \text{ and}$$

for dark rings,

$$2t \cos \phi = (2n-1)\,\frac{\lambda}{2}$$

For *normal* incidence,

$$\phi = 0$$

$$2t = n\lambda \text{ for a bright ring}$$

and

$$2t = (2n-1)\,\frac{\lambda}{2} \text{ for a dark ring.}$$

Proceeding in the same manner as in the reflected system, we can express the diameters of the rings in terms of R and λ.

For nth bright ring,

$$d_n^2 = 4nR\lambda, \quad ...15$$

and for nth dark ring,

$$d_n^2 = 2(2n - 1) R\lambda. \quad ...16$$

where symbols have their usual meanings.

On comparing the corresponding equations in the two systems – reflected and transmitted— it becomes evident that a dark ring in the reflected system becomes a bright ring in the transmitted system and *vice versa*. The central spot which is dark in the reflected system is bright in the transmitted system. Thus the two systems are exactly complementary to each other. But they *differ* in one important respect. In the reflected system, the intensities of the two interfering beams are almost equal so the contrast between the bright and dark rings is considerable, whereas in the transmitted system the contrast between bright and dark rings is small. This is due to the fact that the amplitude of the ray transmitted directly through the air film is so much greater than that which is twice internally reflected in the air film that there is not much difference between the intensities of the bright and dark rings.

Example. *Newton's rings are formed between a plane surface of a glass and a lens. The diameter of the third dark ring is 3.2 mm, when a light of wavelength 5890×10^{-8} cm is used at such an angle that the light passes through the air film at an angle of 30° to the normal. Find the radius of the surface of the lens in contact with glass plate.*

Here, $n = 3$, $\lambda = 5890 \times 10^{-9}$ cm ; $d = 0.32$ cm ; $\phi = 30°$.

For the dark ring using the formulae

$$2t \cos \phi = n\lambda$$

and $$\frac{d_n^2}{4R} = 2t, \text{ we have}$$

$$\frac{d_n^2}{4R} \cos\phi = n\lambda.$$

$$\therefore \quad R = \frac{d_n^2 \cos\phi}{4n\lambda}$$

$$= \frac{(0.32 \text{ cm})^2 \times \cos 30°}{4 \times 3 \times 5890 \times 10^{-8} \text{ cm}}$$

$$= \frac{(0.32 \text{ cm})^2 \times 0.8660}{12 \times 5890} \times 10^{-8} \text{ cm}$$

$$= \mathbf{125.4 \text{ cm} = 1.254 \text{ m}.}$$

Example. *Newton's rings are formed by reflection between two plano-convex lenses with convex surfaces facing each other. The radii of curvature of the convex surfaces are equal being 1 meter each. Find the distance between the 10th and the 20th dark rings using monochromatic light of wavelength* 6×10^{-6} *cm.*

Applying the theory of Newton's rings, we can easily calculate that the thickness of the air film t in this case is given by

$$t = t_1 + t_2 = \left[\frac{d^2}{8R_1} + \frac{d^2}{8R_2}\right]$$

where d and R have their usual meanings.

For the nth dark ring when the incidence is almost normal

$$2t = n\lambda.$$

or $$2\left[\frac{d_n^2}{8R_1} + \frac{d_n^2}{8R_2}\right] = n\lambda,$$

$$\text{or} \quad \frac{d_n^2}{4}\left[\frac{1}{R_1}+\frac{1}{R_2}\right] = n\lambda,$$

$$\text{or} \qquad \frac{d_n^2}{4} = \frac{n\lambda R_1 R_2}{(R_1+R_2)},$$

$$\text{or} \qquad r_n^{\,2} = \frac{n\lambda R_1 R_2}{(R_1+R_2)},$$

$$\text{or} \qquad r_n = \sqrt{\frac{n\lambda R_1 R_2}{(R_1+R_2)}}$$

where $r_n \left(=\frac{d_n}{2}\right)$ is the radius of the nth dark ring.

Here, $\lambda = 6 \times 10^{-5}$ cm ; $R_1 = R_2 = 100$ cm ; $n_1 = 10$, $n_2 = 20$.

Thus the distance between the 20th and 10th dark rings

$$r_{20} - r_{10} = \sqrt{\frac{20 \times 6 \times 10^{-6}\text{ cm} \times 100\text{ cm} \times 100\text{ cm}}{200\text{ cm}}}$$

$$-\sqrt{\frac{10 \times 6 \times 10^{-5}\text{ cm} \times 100\text{ cm} \times 100\text{ cm}}{200\text{ cm}}}$$

$$= -\sqrt{\frac{6}{100}}\text{ cm} - \sqrt{\frac{3}{100}}\text{ cm} = \frac{\sqrt{6}-\sqrt{3}}{100}\text{ cm}$$

$$= \mathbf{0.07174\ cm}$$

Measurement of wavelength of monochromatic light by Newton's rings. The apparatus for studying Newton's rings depicted is used for finding the wavelength of monochromatic light. To perform the experiment place the given source of monochromatic light at S. Clean the Newton's rings device consisting of lens L_2 and glass plate P with petrol and make the necessary adjustments to observe the Newton's rings. Focus the low power travelling microscope on the rings thus formed. Measure with its help the diameters of the nth and $(n + m)$th dark rings as counted from the centre. Also find the radius of curvature R of the lower surface of the lens with the help of a spherometer using the relation

$$R = \frac{l^2}{6h} + \frac{h}{2}.$$

Let d_n be the diameter of the nth dark ring and $(n + m)$th dark ring, then we have

$$\frac{d_n^2}{4R} = n\lambda,$$

and $$\frac{d_{n+m}^2}{4R} = (n + m)\lambda.$$

Subtracting, we get

$$\frac{d_{n+m}^2 - d_n^2}{4R} = m\lambda,$$

or $$\lambda = \frac{\left(d_{n+m}^2 - d_n^2\right)}{4Rm} \quad \text{... (17)}$$

Thus all the quantities on the right band side of the equation being known, the wavelength of monochromatic light used can be calculated.

Determination of refractive index of a transparent liquid by Newton's rings. The transparent liquid whose refractive index is to be measured is introduced between the lens L_2 and the glass plate P and the rings are obtained in the same manner as described above. The diameters of the nth and $(n + m)$th dark rings are then measured with the help of a *low power* microscope. Measuring the radius of curvature R of the *lower* surface of the lens with a spherometer and knowing λ the wavelength of light employed, value of μ for the given liquid is calculated as under:

With the introduction of the liquid film the path difference for the two interfering beams will increase from $2t$ to $2\mu t$.

Now, $$2\mu t = \frac{\mu d_n^2}{4R}, \qquad \text{cf. Equ. (8)}$$

$\therefore$ for the nth dark ring,

$$\frac{\mu d_n^2}{4R} = n\lambda$$

and for the $(n + m)$th dark ring,

$$\frac{\mu d_{n+m}^2}{4R} = (n + m)\,\lambda,$$

Subtracting, we get

$$\frac{\mu\left(d_{n+m}^2 - d_2^n\right)}{4R} = m\lambda,$$

$$\therefore \qquad \mu = \frac{4Rm\lambda}{d_{n+m}^2 - d_n^2}. \qquad \text{...(18)}$$

A similar relation can be deduced if instead of dark rings, measurements are carried out on bright rings.

When the wavelength of light used is not known, then we measure the diameter of a particular dark ring (say nth) first with air and then with the liquid introduced in between the lens and the glass plate, due care being taken not to disturb their relative position. It may be noted that the rings will be closer together with the liquid in between the surfaces.

Suppose d_{air} and d_{liq} are diameters two cases, then

$$\left(d_{air}\right)^2 = 4Rn\lambda$$

and $$\mu(d_{liq})^2 = 4Rn\lambda.$$

$$\therefore \qquad \mu = \left(\frac{d_{air}}{d_{liq}}\right)^2 \qquad \text{... (19)}$$

or better still, we measure the diameters d_n and d_{n+m} of the nth and $(n + m)$th dark rings with air and the diameters d'_n and d'_{n+m} with the liquid introduced in between the lens and the glass plate, then

$$d^2_{n+m} - d^2_n = 4Rm\lambda$$

and also

$$\mu\,(d'^2_{n+m} - d'^2_n) = 4Rm\lambda,$$

$$\text{whence} \qquad \mu = \frac{\left(d^2_{n+m} - d^2_n\right)}{\left(d'^2_{n+m} - d'^2_n\right)} \qquad \text{... (20)}$$

In the discussion of Newton's rings it is assumed that the surfaces are tangent at the point of contact, but actually this is not so. Even if the surfaces are thoroughly cleaned to remove all dust particles and they touch each other, still there will be a slight distortion by the pressure of contact so that the central spot is rarely well-defined. Thus, it is more accurate to measure the diameters of two rings well removed from the centre than the nearer ones. A better method is to take measurements of the diameters of a large number of rings and to plot d^2 against n.

It may be further noted that to produce Newton's rings a point source is *ideal*, for we need strictly parallel light to be incident on the system to give the best definition to the rings. Of course, the rings would be much reduced in brilliancy, and therefore, with a point source it may be difficult at first to observe the rings. So to begin with, it is desirable to employ an extended source when the rings would be very brilliant and a fairly large number would be observed if the microscope were focused on the air film. Want of parallelism in the light, however, makes the rings very broad without suffering displacement and since the air film is of very small thickness a considerable deviation from collimation may be tolerated for a reasonable number of orders to be seen clearly.

With a lens of large curvature, however, the high order rings are quickly lost because the increase in the optical path affects the departure from collimation very seriously, for since $n\lambda = 2\mu t \cos \phi$ for a bright ring in the transmitted system

$$\frac{dn}{d\phi} = -n \tan \phi, \quad \text{... (21)}$$

and of coarse n is proportional to the length of the optical path.

Example. *Newton's rings are formed with reflected light of wavelength 5890 Å using a plano-convex lens and a plane glass plate with a liquid between them. The diameter of the 10th ring is 4.2 mm and the radius of curvature of the curved surface of the lens is 1 meter. Find the refractive index of the liquid if the ring is (i) dark and (ii) bright.*

Here $\lambda = 5890 \times 10^{-8}$ cm ;

$R = 100$ cm ;

$d_{10} = 0.42$ cm.

(*i*) *When the ring is dark,* then

$$\frac{\mu d_n^2}{4R} = n\lambda,$$

or
$$\mu = \frac{4R}{d_n^2} n\lambda$$

$$= \frac{4 \times 100 \text{ cm} \times 10 \times 5890 \times 10^{-8} \text{ cm}}{0.42 \text{ cm} \times 0.42 \text{ cm}}$$

$$= \mathbf{1.335}$$

(*ii*) *When the ring is bright.*

or
$$\frac{\mu d_n^2}{4R} = (2n-1)\frac{\lambda}{2},$$

$$\text{or} \qquad \mu = \frac{4R\left(2n-1\right)\frac{\lambda}{2}}{d_n^2}$$

$$= \frac{4 \times 100 \text{ cm} \times \left(2 \times 10 - 1\right) \times 5890 \times 10^{-8} \text{ cm}}{0.42 \text{ cm} \times 0.42 \text{ cm} \times 2}$$

$$= \mathbf{1.268}.$$

Newton's rings with a bright centre in the reflected light. The fact that the central spot of Newton's rings in the reflected system is dark affords us an experimental proof of the statement that a phase change of π is introduced between the rays reflected from surfaces of denser-to-rarer and rarer-to-denser media. If there were no phase change, the rays reflected from two surfaces in contact should be in the same phase at the centre which should, therefore, be bright.

It is possible, however, to obtain Newton's rings with a bright centre when seen by reflected light. Let us suppose the lower plate P has an index of refraction μ_2 higher than that of the material of the lens μ_1 and a transparent liquid of intermediate refractive index μ is introduced to form a film, such that $\mu_1 < \mu < \mu_2$, a phase change equivalent to an increase in path of $\lambda/2$, where λ is the wavelength of light in the liquid, occurs at both the upper and the lower glass-liquid interfaces; then if the thickness of the film is very small in comparison with the wavelength of light the central spot in the reflected system will be bright. It is because both the reflections are at rarer-to-denser surface and no relative phase change occurs. This experiment was due to Young who observed this change in the central spot by using a *crown* glass lens, a *flint* glass plate and a film of sassafras oil having the value of refractive index midway between the two kinds of glass.

If we can arrange that $\mu_1 < \mu < \mu_2$ the change of phase is zero at each of the boundaries, so that a bright spot will again form the centre of the reflected ring system.

Non-reflecting film (coating of lenses). An important modern application of the phenomenon of interference is the *coating* of optical surfaces with non-reflecting films. This serves to reduce the loss of light by reflections in compound optical systems. At a single reflection about 4% to 5% of the incident light is lost. With a combination of 10 lenses as is usual in compound microscopes, long range finders, periscopes and fast camera lenses about 34% of light is lost because of reflections.

These reflection losses, extremely undesirable specially under conditions of low luminosity, can be avoided by the use of so-called non-reflecting films deposited on the glass surfaces. A thin layer of transparent material of refractive index intermediate between that of air and glass is deposited on the glass surface. Since at both the top and bottom surfaces of the film the reflection is from a denser medium, the same phase change of π occurs at both the reflections, but if the optical thickness of the film μd is made one quarter of a wavelength, then for normal incidence rays 1 and 2 will be exactly π radians out of phase with each other. There is therefore annulment of the two beams and the intensity of the reflected light is a minimum. To obtain zero intensity the two reflected beams should have equal intensities. It can be shown mathematically that this condition can be fulfilled when the index of thin layer,

$$\mu_a = \sqrt{\mu_g},$$

where μ_a is the index of glass.

Magnesium fluoride and crydita (sodium aluminium fluoride) with refractive indices 1.38 and 1.36 respectively when used with heavy flint glass (μ = 1.7) satisfy this condition approximately but when used with crown glass (μ = 1.5) the effect is not so satisfactory. Even when this condition is not fully satisfied the intensity of reflected light approaches zero.

Since the non-reflecting film can have the correct thickness for one wavelength only, a wavelength in yellow-green (5540 A) where the eye has its maximum sensitivity is chosen ; the overall minimum approaches closely to zero, when white light is used.

This treatment of lens and prism surfaces reduces the loss of light by reflection at each surface from about 5% to a fraction of 1 %. So in multi-surface optical systems it is common practice to coat all surfaces to decrease light loss. Also glare, due to reflections from the lens surfaces is so reduced by this coating process that photographs may then be taken with the camera pointed almost directly at the sun.

6

Independent Effect

When two or more sets of waves pass through and cross one another in the *same* medium they behave as if the other were not present, that is, the effects produced by one are totally independent of the effects due to the other. *At any instant the resultant displacement of a particle depends upon the phase difference between the waves and is the algebraic sum of the displacements it would hare at the same instant due to each separate set.* This is known as the principle of superposition and forms the basis of the whole theory of interference discovered by Young in 1801. With a single source of disturbance the distribution of energy in the surrounding medium is uniform but when there are two adjacent sources giving out continuous waves of the same wavelength and amplitude and having the same phase or a constant difference in phase the distribution of energy is no longer uniform.

At some points where the crest of one wave falls upon the trough of the other and *vice-versa,* the resultant amplitude is reduced to zero and the energy is minimum. While at other points where the crest of one wave falls upon the crest of the other or trough of one falls upon the trough of the other the resultant amplitude is increased and the energy becomes maximum. This modification in the distribution of light energy obtained by the superposition of two or more waves is called 'Interference'.

This phenomenon may be effectively illustrated by considering waves on the surface of water or ripples on the surface of mercury. Let us consider two points *A* and *B*, on the surface of water from where two exactly similar circular waves (ripples) are produced with the help of a vibrating tuning fork, to one prong of which two parallel wires are attached and made to dip in water. These circular waves travel out in the form of crests and troughs on the surface of water. Let the continuous lines represent the crests and broken ones represent the troughs. The points where a crest falls upon a crest and a trough on a trough are marked by the crosses (×). These crossed points represent points of maximum energy. Here amplitude is doubled and the energy becomes four times. The lines joining these points represent the regions of maximum flow of energy. While other points, where a crest falls upon a trough and *vice versa*, are marked by circles (O). These are the points of minimum energy, since the resultant amplitude at these points is zero. The lines joining the points marked by circles represent regions of no flow of energy.

It should be carefully noted in this phenomenon there is only transference of energy from one region to the other, the energy *missing at one region reappears at another, so to say there is only redistribution and no destruction or creation of energy as a result of the interference of the two wave trains, total average energy of the medium remaining the same.*

Production of *beats* when two sound waves of nearly the same frequency are trailing along the same line in the same direction is another example of the principle of superposition.

Interference Bands

To find the points of maximum and minimum energy consider two sources of light A and B, lying very close to each other and emitting out continuously waves of monochromatic light of wavelength A and of the same amplitude ana in the same phase. At some point O equidistant from A and B, the two wave-trains always arrive in the same phase and reinforce each other producing maximum intensity. At some other point P where AP and BP are not equal there will be a difference in phase which will depend upon the path difference ($BP - AP$). If this is an even multiple of $\lambda/2$, the waves arrive at P in the same phase and reinforce each other producing maximum intensity. When, however, ($BB—AP$) is an odd multiple of $\lambda/2$ the two waves arriving at P are in opposite phase, they cancel out each other and minimum intensity results.

Putting it in mathematical form, we have

For maximum intensity at P, $(BP—AP)=2n\frac{\lambda}{2}$.

When $n = 0$, we get central bright band at O and when $n = 1, 2, 3$, etc., we get the first, the second, the third, etc. bright bands on either side of O.

For minimum intensity at P, $(BP - AP) = (2n—1)\ \frac{\lambda}{2}$.

When $n = 1, 2, 3$, etc., we get the positions of various dark bands.

Thus, *the central bright band is flanked on both sides by bands, which are alternately of minimum ana maximum intensity.* These bands -are called interference bands.

Conditions for permanent interference of light waves. Coherent sources. To observe the phenomenon of sustained or continuous interference of light waves at a place the following conditions should be satisfied :

The two sources should continuously emit waves of the same wavelength or frequency, since whatever we observe is the result of millions of waves so that the resultant displacement will be constant only if the periodic time is the same.

For obtaining interference fringes, the amplitudes of the two interfering wave trains should be equal or very nearly equal. This will reduce general illumination and will produce complete darkness at the positions of minimum intensity.

The two sets of interfering wave trains should either have the same phase or a constant difference in phase that is, they must originate from a single parent source of light. The phase difference of the waves arriving at any point on the screen will depend upon *(i)* the path difference of the point from the two sources and *(ii)* the phase relation between the waves at the time of emission from the source. The former for the given point remains the same depending as it does on the wave length which is constant, while the latter rapidly changes with time not *only* in different sources but even in *different* parts of the same source.

To understand this point let us go back to the origin of light which is ultimately tithable to the atoms. According to the accepted theory of atomic structure, an atom consists of a positively charged nucleus with electrons revolving in definite orbits around it. When the atom is excited, say, by heating or electric discharge, etc., then the revolving electrons absorb energy and go to outer orbits which are unstable. Hence soon the electrons spontaneously fall back to inner orbits thus emitting energy in the form of radiation. When the frequency of these radiations lies within the range of visible spectrum, we have the emission of light, which consists of a broken chain of wave-trains, accompanied with sudden and abrupt changes in phase occurring in very short intervals of time (of the order of 10^{-8} sec.). Consequently with two independent sources or with two separate portions of the same source, the phases of waves emanating from them will be changing independently of each other so that the phase difference between two such sources cannot be constant. This results in rapid alternations of brightness and darkness and hence in general illumination. This explains, why it is not possible to produce interference with two independent sources, such as two lamp filaments set side by side. On the other hand, in the case of a real source and its image or two images of the same source, any change of phase in the real source will cause a simultaneous and equal change in any image of it which may be produced. Thus it is very essential to have two narrow sources which satisfy the conditions 1, 2 and 3 above. Such sources are known as coherent sources They can be obtained either by (*i*) the source and its *virtual* image or (*ii*) the two *virtual* images of the same source. It is also possible to split up the beam coming from a single source into two beams as if coming from two sources.

The two sources should be very narrow. A broad source is equivalent to a large number of narrow sources lying side by side. Each set of these sources will produce an interference pattern of its own which will overlap one another to such an extent that all traces of a fringe system is lost and general illumination results.

The two sources emitting a set of interfering beams should be very close to each other, that is, the two interfering wavefronts must intersect at a very small angle. Otherwise, since the wavelength of light is extremely small, the path difference between the interfering wave trains will be large and the interference bands will be very close to each other and will not be sharp. They may even overlap each other.

For complete annulment or positions of minima the phase difference between the waves from the two sources should be π or $(2n + 1)\pi$ and for the positions of maxima (reinforcement) it should be 2π or $2n\pi$.

Methods of producing interference. The instruments producing interference phenomena may be grouped into two classes :

(*A*) those which affect division of wavefront into two parts by reflection, refraction or diffraction so that afterwards those two parts reunite at a small angle and produce interference ; and

(*B*) those which affect the division of amplitude of a section of the wave front into two parts by a combination of reflection and refraction to be reunited later to produce interference.

In every instrument of class *A* it is necessary either *to* use a point source or a narrow illuminated slit parallel to the intersection of the two wavefronts. In these devices diffraction effects are also observed on account of a limited portion of the wavefront being used but they can easily be distinguished from interference effects as we shall explain later. The Young's slits, Fresnel's bi-prism and mirrors and Lloyd's mirror described in this chapter and Rayleigh's refractometer described in a subsequent chapter belong to this class.

In the instruments of class B it is not essential to employ a point source or a narrow line source but an extended bright source is invariably used to obtain bright fringes. Since in this case the wavefront is divided in amplitude, if upon reunion corresponding points in the separate parts are superposed the condition of coherence of sources is satisfied. The thin film devices, Newton's rings and Michael son's and several other interferometers etc , belong to this class.

Another point of distinction may also be noted in the two classes of instruments mentioned above depending upon the existence or non-existence of the *complementary pattern.* In general, devices of class A *(Division of* Wavefront*)* do not produce complementary patterns, while those of class B

(Division of Amplitude) with few exceptions do produce them (see Newton's rings).

Division of Wavefront

Young's experiment on interference. The phenomenon of interference was first observed by an English scientist Thomas Young (1773-1829), who successfully demonstrated it in 1802; thus definitely establishing the wave-character of light.

Young allowed a beam of sunlight to pass through a pin-hole S, into a darkened room and then at some distance away, through two pin-holes S_1 and S_2 placed sufficiently close together and symmetrically with respect to S. Here S_1 and S_2 serve as two *coherent* sources of light of the *same* intensity. The two sets of spherical waves coming out of these two pin-holes interfered with each other in such a way as to form a symmetrical pattern of varying intensity on the screen XY.

If the pin-holes be replaced by narrow slits, held perpendicular to the plane of the paper, and illuminated by monochromatic light, there will be cylindrical wavefronts in place of spherical wavefronts. A close examination of the light on the screen will show *evenly spaced* bright and dark fringes parallel to the length of the slits in the direction at right angles to S_1S_2. These fringes diminish in width as the distance S_1S_2 is increased until finally they disappear Young's experiment can be easily demonstrated in the laboratory.

To get widely-spaced fringes the slits should be *quite close* to each other, about 0'2 mm apart. Monochromatic light can be obtained with the help of coloured glass filters The red waves due to their longer wavelength produce wider fringes than those due to the green of shorter wavelength.

Theory of Interference Fringes

To find the positions at which these bands are formed and

to investigate their spacing and the intensity at any point P, let us assume that the two slits S_1 and S_2 are equidistant from the source slit S, so that the light vibrations at the slits will be in the same phase at any instant.

Let $S_1S_1 = d$, $OC=D$ and $CP = x$. With P as centre and PS_1 as radius A, so that S_2A ($= S_2 P - S_1P$) draw an are cutting PS_2 at A, so that S_2A ($= S_2P - S_1 P$) is the path difference between the two wave-trains starting from S_1, S_2 and reaching P.

Now, $$S_2P^2 = D^2 + \left(x + \frac{d}{2}\right)^2$$

and $$S^1P^2 = D^2 + \left(x - \frac{d}{2}\right)^2$$

$$\therefore \quad S_2P^2 - S_1P^2 = \left[D^2 + \left(x \frac{d}{2}\right)^2\right] - \left[D^2 + \left(x - \frac{d}{2}\right)^2\right]$$

$$= \left(x + \frac{d}{2}\right)^2 - \left(x - \frac{d}{2}\right)^2 = 2xd,$$

or $(S_2P - S_1 P)(S_2P + S_1P) = 2xd$

or the *path difference*

$$S_2 P - S_1P = \frac{2xd}{(S_2P + S_1P)} = \frac{2xd}{2D} = \frac{xd}{D},$$

since$(S_2 P + S_1P) = 2D$ approximately.

And the corresponding *phase difference*

$$d = \frac{2\pi}{\lambda} \times \frac{xd}{D}$$

Bright fringes. If the path difference, $(S_2P - S_1P)\left(= \frac{xd}{D}\right)$ is an even multiple of $\lambda/2$, the point P is bright

$$\therefore \quad \frac{xd}{D} = 2n\,\frac{\lambda}{2}$$

or $$x = \frac{D}{d} n\lambda, \qquad ...(1)$$

gives the condition for a *bright fringe.*

Dark fringes. If the path difference is an odd multiple of $A/2$, the point P is dark.

$$\therefore \qquad \frac{xd}{D} = (2n - 1) \frac{\lambda}{2},$$

or $$x = \frac{D}{d} (2n - 1) \lambda/2 \qquad ...(2)$$

gives the condition for a *dark fringe.*

Thus we get an alternate bright and dark system of short, straight and practically parallel fringes on the screen. These fringes are called interference fringes or interference bands.

The whole number n which characterizes a particular bright fringe denotes the order of fringe system. Thus the bright fringes with $n = 0, 1, 2, 3$......are called the central, the first, the second, the third etc., order fringes. Their positions are given by

$$x = 0, x_1 = \frac{D}{d} \lambda, x_2 = 2\lambda, x_3 = \frac{D}{d} 3\lambda,$$

....

$$x_n = \frac{D}{d} n\lambda, x_{n+1} = \frac{D}{d} (n + 1) \lambda.$$

The distance between two consecutive bright fringes is given by

$$x_{n+1} - xn = \frac{D}{d} (n + 1) \lambda - \frac{D}{d} - n\lambda = \frac{D}{d} \lambda.$$

Now for dark fringes

$$x = \frac{D}{d} (2n - 1) \frac{\lambda}{2},$$

where $n = 1, 2, 3,$ for the first, the second and the third order fringes and so on.

So the positions of the dark fringes are given by

$$x_1 = \frac{D}{d} \cdot \frac{\lambda}{2}, \; x_2 = \frac{D}{d} \cdot \frac{3\lambda}{2}, \; x_3 = \frac{D}{d} \cdot \frac{5\lambda}{2},$$

$$\ldots \quad \ldots \quad \ldots \quad \ldots \quad \ldots \quad \ldots$$

$$x_n = \frac{D}{d} (2n - 1) \frac{\lambda}{2}, \; x_{n+1} = \frac{D}{d} (2n + 1) \frac{\lambda}{2}$$

The distance between the two successive dark fringes is given by

$$x^{n+1} - x_n = \frac{D}{d} (2n+1) \frac{\lambda}{2} - \frac{D}{d} (2n-1) \frac{\lambda}{2} = \frac{D}{d} \lambda.$$

Thus the distance between the two successive dark and bright fringes is the same. This distance is called fringe width and is independent of the order of the fringe. Let it be denoted by β

$$\therefore \qquad b = \frac{D}{d} \lambda. \qquad \ldots(3)$$

That is, the fringe-width β is *directly* proportional to D and λ, and *inversely* proportional to d.

If instead of slits S_1 S_2 we have two point sources, even then fringe system will consist of short narrow practically parallel straight bands. To explain this let us consider the effect at some point Q on the screen where, PQ is at right angles to PC. Let PQ be equal to y, then

$$S_2Q^2 = S_2P^2 + y^2 = D^2 + \left(x + \frac{d}{2}\right)^2 + y^2$$

and $$S_1Q^2 = S_1P^2 + y^2 = D^2 + \left(x - \frac{d}{2}\right)^2 + y^2$$

$$\therefore \qquad S_2Q^2 - S_1Q^2 = S_2P^2 - S_1 P^2 = 2xd,$$

or $$(S_2 Q - S_1Q)(S_2Q + S_1Q) = 2xd.$$

Thus, if PQ be small we may take $S_2Q + S_1 Q = 2D$ approximately, giving

$$(S_2Q - S_1Q) = \frac{2xd}{2D} = \frac{xd}{D} \text{ as before.}$$

So, if there is brightness at P, there is brightness at points on the line on the screen through P cutting CP at right angles.

It is obvious from equation (3) that a measurement of the fringe-width gives a direct determination of the wavelength of light in terms of known quantities.

Example. *Green light of wavelength 5100 A from a narrow slit is incident on a double slit. If the over all separation of 10 fringes on a screen 2 m away is 2.0 cm, find the double slit separation.*

Here $\lambda = 5100 \times 10^{-8}$ cm ; $D = 200$ cm ; $\beta = \frac{2.0}{10} = 0.2$ cm.

Let the double-slit separation be equal to d cm.

Using the relation, $\beta = \frac{D}{d}\lambda$, we have

$$d = \frac{D}{\beta}\lambda = \frac{200\text{cm}}{0.2\text{cm}} \times 5100 \times 10^{-8} \text{ cm.}$$

$$= 0.051 \text{ cm} = \mathbf{0.41\ mm.}$$

Example. *Two narrow parallel slits 0 5 mm apart are illuminated by a monochromatic light of wavelength 5890 A. Calculate the width of the fringes which are obtained on a screen distant 50 cm from the slit.*

Here, $d = 0.5$ mm $= 0.05$ cm; $\lambda = 5890 \times 10^{-8}$ cm ;

and $D = 50.0$ cm.

Using the relation, $\beta = \frac{D}{d}\lambda$, we have

the fringe-width, $\beta = \frac{50.0 \text{ cm}}{0.05 \text{ cm}} \times 5890 \times 10^{-8} \text{ cm} = 0.0589 \text{ cm.}$

Distribution of intensity in the fringe system. It has been discussed, that when the path difference $(S_2P - S_1P)$, *i.e.*, S_2A, contains a whole number of wavelengths or an even multiple of half wavelengths, the two waves reach P in the same phase and reinforce each other so that the resultant amplitude is equal to the sum of the separate amplitudes. On the other hand, if S_2A contains an odd multiple of half wavelengths, the two waves reach P exactly in opposite phase and the resultant amplitude is their difference. Maximum intensity occurs at C which is equidistant from S_1 and S_2 and also at P_1 P', P_2, P'', *etc.*, because each of these points is some whole number of wavelengths further from one slit than from the other. If the screen is moved towards or away from the slit the spacings of these maxima decrease or increase nearly in direct proportion of the slit-screen distance D, since $\beta = \frac{D}{d}\lambda$. It may be noted that the fringes are not straight as shown by the elementary theory given above but actually they are parts of *hyperbolas*, as it is a hyperbola which is the locus of a point for which the difference in the distance *from two fixed points is constant*. In this case, however, the hyperbolas are producing short straight lines as the wavelength is small and the distance of the screen large.

To find the intensity on the screen due to two interfering beams,

$$y_1 = a_1 \sin \frac{2\pi}{\lambda} (vt - x), \text{ and}$$

$$y_2 = a_2 \sin \frac{2\pi}{\lambda} [(vt - x) + \delta)]$$

where a_1 and a_2 are the amplitudes of the interfering beams, δ the phase difference, then the resultant amplitude is given by

$$R^2 = a_1^2 + a_2^2 + 2a_1 a_2 \cos \delta,$$

from which it is clear that for a maximum, $\delta = 0$ and $R = a_1$

$+ a_2$; and for a minimum, $\delta = 0$, *i.e.*, a_1 and a_2 are parallel and in opposite direction so that $R = a_1 - a_2$. In general for any value of δ, R is the closing side of the triangle OAB as shown in R^2 measures the intensity at a point which is given by

$$I = R^2 = 2a^2 (1 + \cos \delta), \text{ if } a_1 = a_2 = a, \text{ say}$$

Then $$I = 4a^2 \cos^2 (\delta/2). \qquad ...(4)$$

The intensity distribution in the interference pattern which is proportional to $\cos^2 (\delta/2)$ is shown in the curve where the values of δ *viz.*, $0, \pm \pi, \pm 2\pi, \pm 3$ etc. are marked at equidistant points on the axis of X and those of I along the axis of Y through $\delta = 0$

For $\delta = 0, \pm 2\pi, \pm 4\pi, \pm 6\pi,$

etc. $I = 4a^2$

and for $\delta = \pm \pi, \pm 3\pi, \pm 5\pi$ etc. ; $I = 0$.

Thus the intensity varies between zero and $4a^2$ according as the phase difference δ between the interfering beams is odd π or even π.

There is no *destruction of energy* in the interference phenomenon. The energy which apparently disappears at the minima, has actually been transferred to the maxima where the intensity is greater than that produced by the two beams acting separately. Each beam acting separately contributes an intensity equal to a^2 and so without interference we will have a uniform intensity of $(a^2 + a^2) = 2a^2$ as shown by the horizontal dotted lino which gives the average value of the intensity on the screen. This result is obtained by inserting in equation (4) the mean value of $\cos^2 (\delta/2)$ between 0 and π which is given by

$$\frac{1}{\pi}\int_0^\pi \cos^2 \frac{\delta}{2}\, d\delta = \frac{1}{2\pi}\int_0^\pi (1 + \cos\delta)\, d\delta = \frac{1}{2}$$

$$\ldots \left[\because \int_0^\pi \cos\delta\, d\delta = 0\right]$$

Thus the energy is not destroyed but merely redistributed bated in the interference pattern. Hence the interference phenomenon is in conformity with the law of conservation of energy.

Fresnel's Experiments

Even through Young had demonstrated the interference phenomenon quite successfully by his experiments, still there were doubts and it was thought that the fringes observed by Young might probably be due to some complicated modification of light by the edges of the slit and not to true interference. To remove these doubts Fresnel devised several new experiments by obtaining two small sources of light close together wholly independent of aperture or edges of opaque obstacle such as *bi-prism experiment, double mirror experiment etc.*

Bi-prism Experiment

A bi-prism, as its name suggests, is a combination of two thin prisms with their bases joined and their two faces making an obtuse angle of about 179° so that the other angles are each of about 30'. In actual practice the bi-prism is ground from a *single optically true glass* plate. A monochromatic source of light illuminates a narrow vertical slit S held symmetrically and at a short distance from a bi-prism P whose refracting edges are vertical. Each half of the bi-prism produces a virtual image of S by refraction. The distance between S and P is so adjusted that the two *virtual* images S_1, S_2, are quite close together. The diagram represents the horizontal cross-section of the arrangement. Thus Fresnel used the bi-prism to get two *coherent sources,* S_1 and S_2, of monochromatic light from a single source S so that $S_1 S_2 = d$ and the distance of the point C on the screen XY from the source S is D, *i.e.,* $D = SC$.

The bi-prism P refracts light from the vertical slit source S into two overlapping beams AF and BE.

A closely-spaced interference fringe system is produced only in the region *AB* while the wide set of fringes at the edges of the pattern is on account of diffraction.

Since the interference fringes are narrow they are generally examined with the help of a *low-power microscope,* the fringes lying in its focal plane.

Determination of Wavelength of Monochromatic Light

Theory. The point C on the screen is equidistant from the' two virtual sources S_1 and S_2, hence the two waves reaching *C* reinforce each other and so this point will be the centre of a bright fringe, while the illumination at any other point on the screen will depend upon the difference between the optical paths of the point from S_1 and S_2 *i.e.,* $\frac{xd}{D}$.

For the point to be at the centre of a bright fringe,

$$\frac{d}{D} - x = n\lambda.$$

$$\therefore \qquad \frac{D}{d}\lambda = \frac{x}{n} = \beta \text{ the fringe-width.}$$

Thus, if *D* and *d* are constant, the fringe-width β is proportional to wavelength of monochromatic light λ. Hence shorter wavelength fringes are narrower than the longer wavelength fringes. The knowledge of *D* and *d*, and the measurement of the fringe-width in a bi-prism experiment enable us to determine the wavelength λ of monochromatic light.

$$\text{Since} \qquad b = \frac{D}{d}\lambda,$$

$$\therefore \qquad 1 = \frac{d}{D}\beta = \frac{d}{a+b}\beta \qquad ...(5)$$

$$\text{where} \qquad D = SC = SO + OC = a + b.$$

So that knowing *d*, (*a*+*b*) and β, λ can be calculated.

Experimental procedure. A specially designed variable vertical slit *S*, the bi-prism *P* and a micrometer eye-piece *E*, are mounted in uprights capable of vertical ana transverse adjustments and set about 175 cm long with a scale graduated in mm. The various devices are arranged at the same height above the bed of the optical bench. The slit is made narrow and is illuminated by the monochromatic light of which the wavelength is to be determined. The bi-prism is moved at right angles to the bed, so that on looking through it two equally bright victual slits, S_1 and S_2, are observed. Next, the eye-piece which is capable of lateral displacement alongside a horizontal scale is adjusted so that the overlapping region *AB* is brought in its field of view and the fringes are obtained in its *focal* plane.

To make the fringes *well-defined* the slit *S* is rotated in its own plane till it is exactly parallel to the refracting edge of the bi-prism. Finally, it is seen that on moving the micrometer eye-piece no lateral shift of the fringe system is observed. This will occur when the line joining the slit and the edge of the bi-prism is parallel to the bed of the bench. After these adjustments the necessary measurements can be easily made.

(i) *Measurement of fringe width β.* The centre of the particular bright fringe is brought on the cross-wires of the micrometer eye-piece by the slow motion screw attached to it and the position of the eye-pieces is read on the scale. It is moved to one side through a known distance and the number of fringes passing across the view is counted. The total distance moved through by the micrometer divided by the number of fringes gives the mean value of β.

(ii) *Measuring distance (d) between the virtual sources.* To measure *d*, we use the displacement method. A convex lens of *suitable* focal length is placed in between the eye-piece and the bi-prism without disturbing their positions. The distance between a sharp pair of images of the two virtual sources S_1 S_2 for the two conjugate positions of the lens, L_1 and L_2, are measured. Let these distances be x and y respectively. In these positions the magnifications are

$$m_1 = \frac{x}{d} = \frac{v}{u},$$

and
$$m_2 = \frac{y}{d} = \frac{v_1}{u_1} = \frac{u}{v},$$

where u and v are the distances of the object and the image respectively from the lens in the first conjugate position.

$$\therefore \quad \frac{x}{d} \times \frac{y}{d} = \frac{v}{u} \times \frac{u}{v} = 1,$$

or
$$xy = d^2, \quad \therefore \quad d = \sqrt{xy}. \qquad ...(i)$$

(iii) *Measurement of D*. The distance D between the slit S and *the focal plane* of the eye-piece is measured with the help of the optical bench scale. The distance between the indices attached to the slit and eye-piece uprights is read and corrected for index error.

The index correction between the slit and the eye-piece uprights is determined by making one end of a knitting needle touch the slit and focusing the eye-piece at the other end. Then the difference between the length x of the needle and the observed distance y between the uprights *i.e*, $(x - y)$ gives the required index correction, which added to the observed distance between S and E gives the correct value of D.

Thus measuring β, d and D ($= a + b$) and substituting their values in the relation.

$$\lambda = \frac{d}{D}\beta,$$

the wavelength λ can be calculated.

Note. There is an *alternative* method of measuring d which makes use of the fact that for a prism of very small angle, the deviation θ of the ray is given by $\theta = (\mu, - 1)\alpha$, where μ is the refractive index of the material of the prism and α its refracting angle. Both μ and α can be measured with the help of a spectrometer and so θ is found out.

We find that

$$d = 2\theta \times a = 2a\ (\mu - 1)\alpha \qquad ...(ii)$$

$$\therefore \qquad \lambda = \frac{d}{D}\beta = \frac{2a(\mu - 1)\alpha}{D}\beta \qquad ...(6)$$

Example. *In an experiment with Fresnel's bi-prism sodium light is used and bands 0.0196 cm in width are observed at a distance of one meter from the slit. A convex lens is then put between the observer and the prism so as to give image of the sources at a distance of one meter from the slit. The distance apart of the images is found to be 0'70 cm, the lens being 30 cm from the slit. Calculate the wavelength of sodium light.*

According to equation (5) on page 244, $\lambda = \dfrac{d}{a+b}\beta$.

Here β = 0. 0196 cm ; a+6 = I00 cm and the distance apart of the images, x = 0.70 cm.

Since the distance of the lens from the slit to obtain the image of the virtual sources is 30 cm, we have

$$u = 30 \text{ cm}, v = 100 - 30 = 70 \text{ cm}$$

and $$x = 0.70 \text{ cm}.$$

Since $$\frac{I}{O} = \frac{v}{u} = \frac{x}{d},$$

$$\therefore \qquad \frac{70}{30} = \frac{0.70}{d},$$

or $$d = 0.70 \times \frac{30}{70} = 0.30 \text{ cm}.$$

$$\therefore \qquad \lambda = \frac{d}{a+b}\beta = \frac{0.30\text{ cm}}{100\text{ cm}} \times 0.0196$$

$$= \mathbf{5880 \times 10^{-8}\ cm.}$$

Example. *A Fresnel's bi-prism with refracting angle 1° and refractive index 1.55 is used to form interference fringes. Calculate the fringe-width for a light of wavelength 6000 A, when the distance between the source and the prism is 30 cm and that between the prism and the screen 70 cm.*

Here P is the bi-prism, S the source of light, and, S_1 and S_2 the two coherent virtual sources.

Let the distance between the bi-prism and the source of light be represented by a and that between the bi-prism and the screen by b. Suppose the distance between the virtual sources S_1 and S_2 is d.

To find d we make use of the fact that for a prism of very small angle, the deviation of the ray is given by $(\mu - 1)\alpha$, μ being the refractive index of the material of the prism and a its refracting angle.

$$\therefore \qquad \theta = (\mu - 1)\,\alpha.$$

$$\text{Also} \qquad \frac{d}{2} = a\,\theta. \qquad \left[\because \frac{S_1 S_2}{SO} = 2\,\theta \text{ or } \frac{d}{a} = 2\theta.\right.$$

$$\therefore \qquad d = 2a\,(\mu - 1)\,\alpha,$$

$$\text{Now,} \qquad a = 30\text{ cm}\ ;\ \mu = 1.55\ ;\ \alpha = 1^\circ = \frac{\pi}{180}.$$

$$\therefore \qquad d = 2 \times 30\,(1.55 - 1)\,\frac{\pi}{180} = \frac{\pi}{3} - 0.55\text{ cm.}$$

$$\text{Now} \qquad 1 = \frac{d}{a+b}\beta, \text{ where } \beta \text{ is the fringe-width}$$

$$\therefore \qquad \beta = \frac{a+b}{d}\lambda.$$

Putting values of a=30 cm ; b = 70 cm ; A =6000 × 10^{-8} cm

and $d = \frac{\pi}{3} \times 0.55$ cm in this relation, we get

$$\beta = \frac{(30+70)\text{ cm}}{\frac{p}{3} \times 0.55 \text{ cm}} \times 6000 \times 10^{-8}\text{cm.}$$

$$= 0.0104 \text{ cm} = \mathbf{0.104 \text{ mm}}.$$

Fresnel's double mirror experiment. Fresnel used two plane mirrors MM_1 and MM_2, to produce interference fringes. He placed them close together so that their planes make a very small angle θ with each other and intersect in a line perpendicular to the plane of the paper at *M*. A beam of light from a narrow slit *S* placed parallel to the line of intersection of the mirrors and illuminated by a monochromatic source of light is made to fall upon the mirrors. This light is split up into two beams after reflection from the mirrors.

The light arriving at the screen *XY* appears to come from two coherent virtual sources S_1 and S_2. It should be noted that *the re' flection takes place at the front surfaces of the mirrors and not from their back surfaces.*

For this purpose the mirrors should preferably be made of optically flat glass silvered on the front surface or else optically flat black glass should be used.

In the region *BC* on the screen where the two beams *AD* and *CB* respectively from the two *virtual* sources S_1, and S_2 overlap and interfere, a large number of dark and bright *equi-spaced* fringes are seen. The fringes are examined with the help of a micrometer eyepiece and their fringe-width is measured.

Determination of Wavelength of Monochromatic Light

Theory. It follows that $MS_1 = MS_2 = MS$, *i.e.*, *S*, S_1 and S_2 are equidistant from *M*. Let $a(= MO)$ and $b\ (= MP)$ be the distance of the sources and the screen from M and d the distance between the two virtual sources S_1 and S_2. If λ be the wavelength of light used, then the fringe-width β is given by the relation

$$\beta = \frac{D}{d} l$$

$$= \frac{a+b}{d} \lambda .$$

Since MC and MD are the reflected rays of the incident ray SM from the two mirrors MM_1 and MM_2 respectively, $\angle CMD = \angle S_1 MS_2 =$ twice the angle between the mirrors, *i.e.*, 2θ.

$$d = 2\theta \times a,$$

whence $$\beta = \frac{a+b}{2a\theta} \lambda$$

or $$\lambda = \frac{2a\theta}{a+b} \beta . \qquad ...(7)$$

When fringes are obtained with white light, the central fringe is also white, for the intensity is always maximum at P for each wavelength. Since the fringe-width β is a function of wavelength λ the fringes due to different colours overlap resulting in coloured fringes, the dark fringes being only visible in the vicinity of the central white fringe. Due to this overlapping of different fringe systems in the outer parts of the interference field the total number of fringes observed with white light is much smaller than with monochromatic light. The reader should, however, note that overlapping will be sooner in the case of the bi-prism fringes than in the present case on account of the material of the prism being dispersive as described earlier.

Example. *A pair of Fresnel's mirrors making an angle of 2° are held at a distance of one meter from a slit source emitting light of wavelength 5000 A. Find out the fringe-width on a screen distant 2.5 meters from the intersection of the mirrors.*

Here $\theta = 2° = \frac{\pi}{180} \times 2 = \frac{\pi}{90}$ radian ;

$a = 100$ cm ;

$6 = 250$ cm,

$\lambda = 5000 \times 10^{-8}$ cm;

$\beta = ?$

Let d be the distance between the two virtual interfering sources.

Now, $d = 2a\theta = 2 \times 100 \text{ cm} \times \frac{\pi}{90}$ cm.

The fringe separation is given by the relation

$$\beta = \frac{a+b}{d}\lambda$$

$$= \frac{(100+250)\text{ cm}}{\frac{20\pi}{9}\text{ cm}} \times 5000 \times 10^{-8} \text{ cm.}$$

$$= \frac{35 \times 5 \times 9}{2 \times 3.142} \times 10^{-8} \text{ cm.}$$

$$= \frac{35 \times 45 \times 10^{-8}}{6214} \text{ cm}$$

$$= \mathbf{0.0251 \text{ mm.}}$$

Determination of thickness of a thin transparent film by displacement of fringes. We can use bi-prism experiment for determining the thickness of a thin sheet of transparent material such as mica or glass. Let S_1 and S_2, be the two *coherent* monochromatic sources of light which are producing interference fringes on the screen so that C is the position of the central bright band, because the optical paths S_1C and S_2C are equal. When a thin glass plate of thickness t and refractive index μ, (μ >1) is introduced in the path of light from one of the two sources, then due to the difference in the velocity of light in air and glass there is a change in the path difference, so that the fringe system will shift. If the glass plate is placed in the path S_1 P the fringes shift *upward*. The central fringe shifts to the position P_0, so that the optical path S_1P_0 is again equal to the optical path S_2P_0. The point P will be the centre of nth bright fringe if the path difference is equal to $n\lambda$.

Now the path difference between S_2P and S_1P

$$= S_2P - [(S_1P - t) + \mu t]$$

$$= S_2P - S_1P - (\mu - 1)t$$

$$= S_2A + AP - S_1P - (\mu - 1)t$$

$$= S_2A + AP - AP\ (\mu - 1)t$$

$$= S_2A - (\mu - 1)t$$

$$= \frac{xd}{D} - (\mu - 1)t \qquad .. \left[\because S_2A = \frac{xd}{D}\right]$$

where x is the distance of *nth* bright fringe from the central fringe in the absence of the glass plate.

$$\therefore \quad \frac{xd}{D} - (\mu - 1)\ t = n\lambda$$

or

$$x = \frac{D}{d}\left[n\lambda + (\mu - 1)\ t\right] \qquad ...(8)$$

The displacement x_0 of the central bright fringe when the glass plate is placed in the path S_1P is obtained by putting $n = 0$ in equation (8).

$$\therefore \quad x_0 = \frac{D}{d}(\mu - 1)t \qquad ...(9)$$

or

$$t = \frac{x_0\ d}{D(\mu - 1)} \qquad ...\ (10)$$

Thus knowing that D and μ, t may be calculated.

Also since $\mu, > 1$, x_0 is positive. This confirms that the central fringe will shift upward when a thin glass plate is placed in the path. S_2P.

To obtain the positions of the first, the second, the third, etc. bright fringes, we put n=l, 2, 3, etc. in equation (8) and get

$$x_1 = \frac{D}{d} [\lambda + (\mu - 1) t]$$

$$x_2 = \frac{D}{d} [2\lambda + (\mu - 1) t]$$

... ...

Also the fringe-width,

$$\beta = x_{n+1} - x_n$$

$$= \frac{D}{d} [(n + 1)\lambda + (\mu - 1) t] - \frac{D}{d} [n\lambda + (\mu - 1) t]$$

$$= \frac{D}{d} \lambda \qquad ...(11)$$

which is the same for all fringes. This shows that the fringe-width is not affected by the introduction of the glass plate.

From equation (11), we get

$$\frac{D}{d} = \frac{\beta}{\lambda}.$$

Substituting this value in equation (9) above, we have

$$x_0 = \frac{\beta}{\lambda} (\mu - 1)t. \qquad ...(12)$$

The wavelength λ of the light used can be determined if μ, t, x_0 and β are known. The displacement x_0 of the central fringe and the fringe width P can be determined experimentally.

If however, x_0, β, λ and t are given, the value of μ can be calculated.

When monochromatic light is used, all fringes are exactly similar and so it is not possible to distinguish the central bright fringe from the rest of the fringes. Thus the shift of the central fringe with monochromatic light alone cannot be measured. This, however, can be done by using white light. With white light the central fringe is white while all others are coloured, so it can be easily distinguished and correctly located. In actual practice the monochromatic light is superimposed by white light from a frosted lamp. The position of central fringe is noted on the micrometer scale. On introducing the mica sheet in the path of the upper beam, say, the central white fringe shifts upwards. This shift (x_0) can then be accurately measured by the micrometer eye-piece.

It should be clearly noted that the central fringe for white light will be white so long as the glass sheet introduced in the path of light is very thin. Otherwise due to dispersion of light all colours will not be coincident at the central fringe and so it will not be perfectly white.

Example. *A thin film of plastic of refractive index 1.45 for light of wavelength 5890* A *is inverted normally in the path of one of the interfering beams. The central bright band shifts through 5 fringes. Find the thickness of the film.*

Here, $\mu = 1.45$; $\lambda = 5890 \times 10^{-8}$ cm ; $x_0 = 5\beta$, where β is the fringe-width and $t = ?$

Using the relation (11).

$$x_0 = \frac{\beta}{\lambda}(\mu - 1)\, t, \text{ we have}$$

$$5\beta = \frac{\beta}{5890 \times 10^{-8} \text{ cm}}(1.45 - 1)\, t,$$

$$t = \frac{5 \times 589 \times 10}{0.45}$$

$$= \mathbf{6.544 \times 10^{-4} \text{ cm.}}$$

Lloyd's Single Mirror

This apparatus for producing interference fringes was first devised by Dr. H. Lloyd in 1834. It is a single plane mirror and is very simple to adjust. Monochromatic light from a narrow slit source S_1 is made to fall at practically grazing *incidence* on the surface *of a fairly* flat strip of glass *MN* about 30 cm long and several cm wide. The slit is held normal to the plane. To avoid internal reflections the glass plate should preferably be black. The light after reflection appears to come from the virtual source S_2. In addition to the reflected light the screen *AD* also receives light directly from the source S_1 so that S_1 S_2, which are adjusted close together serve as two coherent sources. The two beams overlap in the region *AB* and interfere giving rise to a system of fringes.

The point *C* on the screen lying on the right-bisector of S_1S_2 gives the position of the central fringe corresponding to zero path difference. This central fringe is not visible as no reflected light reaches C. It can, however, be made visible by introducing a thin mica sheet in the path of the direct ray S_1A and so shifting the central fringe towards *A*. If t be the thickness of the mica plate, the path of the direct beam is increased by $(\mu - 1)t$, where μ is the refractive index of mica. So the place of zero path difference is moved upwards by $x_0/\beta = (\mu - 1)t/\lambda$ fringes, *cf.* equation (12).

An important and interesting feature of the experiment lies in the unexpected. It is found that the central fringe and the other fringes which on previous considerations we should expect as bright, are, instead, *dark*. This can easily be seen when the screen is put in contact with the end *N* of the mirror, *i.e.*, in the position *EF*. The edge *N* of the mirror comes at the centre of the dark fringe, that is, the bands are all displaced through half the interval of a band width from the mirror edge. This leads us to the conclusion that the *direct and reflected lights differ in phase by* π. Since the direct beam could not change phase, naturally this phase change of re has occurred in the reflected beam, on reflection at the surface of the *denser* medium.

Thus this experiment in addition to the interference phenomenon shows that the phase of a beam of light is altered by π when it is reflected from the surface of an optically denser medium. If this is the case, then a crest from S_1 will arrive at the same time as a trough from S_2 at the point of zero path difference and a black fringe will be produced. So that the fringes are bright or dark according as the path difference is $(2n + 1)\,\frac{\lambda}{2}$ or $n\lambda$, where n is an integer.

If a white source of light is used in Lloyd's mirror experiment the central dark fringe would be bordered on one side by a few coloured fringes. The overlapping soon increases and the field becomes white.

Determination of wavelength. Lloyd's mirror may be used to measure the wavelength of monochromatic light used. If the screen AD is at a distance D from the source S_1 the fringe-width is $D\lambda/d$, where $d = S_1S_2$ and λ is the wavelength of the light used (*cf.* equation 3). Thus the value of λ may be determined. In performing the experiment the mirror strip is mounted parallel to the bed of an optical bench, the slit being also *parallel* to the bed. The position of the screen is taken by a micrometer eye-piece in whose *focal* plane the fringes are received. The rest of the procedure is the same as in the bi-prism experiment.

Stokes' treatment of phase change on reflection. This phase change on reflection can be explained theoretically by using the *principle of reversibility* of light. Let MN, be the surface of separation of the two media, the lower medium being denser than the upper one. Let a wave of light of amplitude a incident along AB on this surface be partly reflected along BC and partly refracted along BD into the denser medium If r be the co-efficient of reflection, *i.e.*, fraction of the amplitude reflected and t the coefficient of transmission at this surface, the amplitude of the wave along *BC* is ar and that along *BD* is at.

If the reflected and refracted waves *BC* and *BD* are reversed in their paths, the resultant should be the wave of the original amplitude a, provided there is no absorption of energy.

On reversing the reflected ray *BC* we get the refracted component in the more dense medium of amplitude art along *BE*. Similarly, on reversing the refracted ray *BD*, we get the internally reflected component in the more dense medium of amplitude atr' along *BE* where r' is the reflection co-efficient at the surface of the rarer medium. Since the reversal of ar and at, must merely reproduce the amplitude a, the sum of the components along *BE* should be zero so that

$$art + atr' = 0$$

or

$$r = -r'.$$

The *negative* sign in this equation shows that the coefficients of internal and external reflection are numerically equal but of opposite sign But we know from Lloyd's single mirror that there is a change of phase equal to π when light in air is reflected at the boundary of a more dense medium and so this result proves that there is no phase change when light in a more dense medium is reflected at the boundary of air. This also shows that if we have two rays incident on the surface of a medium, one externally at a given angle of incidence and the other internally at the corresponding angle of refraction, *the amplitude of the reflected ray has in the one case the same ratio to the amplitude of the incident ray as in the other but with a change of sign.* There is no change of intensity in the process as it changes from $(ar)^2$ to $(ar')^2$, the amplitude of the wave remaining same numerically.

Lloyd's single mirror achromatic fringes. In the theory of interference we have shown that the width β of the interference fringes is a function of the wavelength of light, since $\beta = D\lambda/$

d. It is for this reason that with white light except the central bright fringe all other fringes are coloured, blue ones being closer together than red ones. The question is, whether we can get achromatic fringes with white light and can make β the same for all wavelengths so that the fringes are successively white and black. This is possible if λ/d in the relation $\beta = D\lambda/d$ is made constant. We can easily realize this condition with Lloyd's single mirror by using a slit illuminated by a narrow spectrum as the source. The narrow spectrum is produced by a diffraction grating with about 200 lines per cm, or a prism of small angle, and the violet end is arranged to be nearer to the front surface of the mirror *M*. The interference will occur between *VR* and its reflected image *VR′* in the front surface of the mirror. The violet fringes are produced by *VV* and the red by *RR′* and so on. Then provided we arrange the distance of the spectrum in such a way that $RR' / VV = \lambda_r\lambda_v$ we shall have λ/d constant for all wavelengths. Since in the spectrum formed by diffraction grating the dispersion is proportional to wavelength, achromatic fringes will result. Fringes are seen in the region of *P* either on a screen or with an eye-piece. A correct adjustment of the distance of *VR* from the surface of the mirror will give us fringes as nearly black and white as possible.

Difference between the Bi-prism and Lloyd's Mirror Fringes

(1) In the case of Lloyd's mirror less than half the field is in view while in the case of Fresnel's bi-prism a large number of fringes on both sides of the central fringe are observed.

(2) In the case of Lloyd's mirror central fringe is dark and is ordinarily not visible, while with Fresnel's bi-prism central fringe is always *bright*.

(3) Another important point of difference is due to the fact that in the bi-prism the coherent sources are obtained

by refraction and in Lloyd's mirror and in Fresnel's double mirror by reflection. In the bi-prism, therefore, the distance, d, between the virtual sources depends upon the refractive index of the material of the prism for the particular light used. Thus the distance between two virtual sources [$d = 2a\ (\mu - 1)$] is a little farther apart for blue than for red rays since $\mu_b > \mu_r$. Hence for the nth bright fringe of the two sets in bi-prism, we get

$$(x_n)_v = \frac{n\lambda_b\,(a + b)}{2a(\mu_b - 1)\alpha} \quad \text{[ef. equation (1) on page 238}$$

Since $\qquad D = (a + b)$

and $\qquad d = 2a\ (\mu - 1)\alpha$

and
$$(x_n)_r = \frac{n\lambda_r\,(a + b)}{2a(\mu_r - 1)\alpha}$$

the subscripts b and r stand for blue and red rays respectively. These equations clearly show that the difference between $(x_n)_b$ and $(x_n)_r$ is greater than if the material of the prism were not dispersive. This results in greater overlapping when white light is used in the bi-prism experiment so that the field rapidly becomes uniformly illuminated.

***Example.** A slit illuminated by sodium light of wavelength 5896 A is held 2'S mm from the plane of a mirror. Find the fringe-width of the bands formed on a screen 1.25 meters from the slit.*

The distance between slit and its *virtual* image in the mirror,

$$d = 2.5 \text{ mm} + 2.5 \text{ mm} = 5 \text{ mm} = 0.5 \text{ cm}.$$

$$D = 125 \text{ cm} ;$$

$$\lambda = 5896 \times 10^{-8} \text{ cm} ;$$

and $\qquad \beta = ?$

Now the fringe-width β is given by the relation,

$$\beta = \frac{D}{d}\lambda = \frac{125\,\text{cm} \times 5 \times 5896 \times 10^{-8}\ \text{cm}}{0.5\,\text{cm}}$$

$$= \mathbf{0.01474\ cm.}$$

Billet's Split-lens

This device for demonstrating interference phenomenon makes use of two halves of a convex lens placed close together to form two real images S_1 and S_2, of the narrow slit S illuminated by a monochromatic source of light. S, and S_2 now act in the same way as the double slit in Young's experiment. The distance between S_1 and S_2 can be changed by adjusting the space between the two halves of the convex lens, a number of interference bands of varying widths can be obtained and observed in the overlapping region *BC*.

Example. *Interference fringes of wavelength 6000 A are formed by Billet's split lens. The distance from the source S to the lens is 20 cm and the focal length of the lens is 10 cm. Calculate the fringe width if the lens halves are separated by 01 mm and the screen is held at a distance of 2 meters from the source.*

To find the position of the interfering sources S_1 and S_2 with respect to the lens, we make use of the relation

$$\frac{1}{v} - \frac{1}{u} = \frac{1}{f}$$

Here $u = -20$ cm :

$f = 10$ cm ;

$v = ?$

$$\therefore \quad \frac{1}{v} + \frac{1}{20} = \frac{1}{10}$$

or $v = 20$ cm.

The value of d, the distance between S_1 and S_2, is given by the relation;

$$\frac{d}{x} = \frac{u+v}{u},$$

or
$$d = \frac{u+v}{u} x,$$

where x, the separation of the two halves of the lens is 0 01 cm.

$$\therefore \quad d = \frac{20+20}{20} \times 0.01 \text{ cm} = 0.02 \text{ cm}.$$

Now the fringe-width β is given by the relation

$$\beta = \frac{D}{d} \lambda$$

Here,
$$D = 200 \text{ cm} - (u + v)$$
$$= 200 \text{ cm} - (20 + 20) \text{ cm} = 160 \text{ cm}$$

Also
$$d = 0.02 \text{ cm and } \lambda = 6000 \times 10^{-8} \text{ cm}.$$

$$\therefore \quad b = \frac{160 \text{ cm}}{0.02 \text{ cm}} \times 6000 \times 10^{-8} \text{ cm} = 0.48 \text{ cm}.$$

The bi-plates. The apparatus consists of two glass plates P_1 and P_1 cut from the same piece and inclined to each other at an obtuse angle. The light from a luminous source S placed on the bisector of the angle between the plates after refraction through them appears to diverge from two identical virtual sources S_1 and S_2. The emergent beams are allowed to fall on a convex lens L which produces two real images A and B, the so-called coherent sources. After diverging from A and B the beams will overlap and produce interference bands.

7

Light Propagation

Propagation of Energy

From the experiments that light travels with a finite velocity (3×10^8 m/sec.), produces heating effect, affects the photographic plates and influences the growth of plants etc., it becomes evident that light, like heat, is a form of energy. The association of light with energy is also confirmed by the fact that it exerts pressure upon the surface on which it falls. Maxwell predicted it on theoretical grounds and Lebedew verified it experimentally in 1901. It has been clearly established by the experiments of Nichols and Hull that the energy contained in unit volume of the incident beam is equal to the pressure of light in dynes per cm^2. That light exerts pressure is also borne out by the fact that the tails of comets are always directed away from the sun. It is because the sun's rays exert pressure. Since light is a form of energy, the problem of its transmission from one place to another is purely a problem of transmission of energy. There seem to be only two modes of propagation of energy.

1. The simplest one is to suppose that the energy is carried by streams of small material particles travelling with finite velocity, and possessing kinetic energy but *not* requiring any material medium for their propagation. This is the basic principle of Newton's emission or corpuscular theory of light.

2. The other method suggests the transference of light energy by wave motion without the actual travelling of the matter. This process requires presence of an intervening medium as a vehicle for the transmission of energy. This gave rise to the wave theory of light which was put forward by Huygens in 1678.

The emission or corpuscular theory. This theory was due to Newton, the famous English philosopher (1642-1727). According to him a luminous body emitted very *minute* and weightless particles, called corpuscles, which travel through empty space (a homogeneous medium) in straight lines in all directions with the speed of light (3×10^8 m/sec.) and carry with them their kinetic energy. The particles move in straight lines simply due to property of *inertia.* When they fall upon the eye they excite the sensation of sight. Different colours of light were associated with *size* of the corpuscles. These corpuscles are considered so small that they can readily pass through the interspaces of the particles of matter. On account of their high speed and practically zero mass they are unaffected in their linear path by the force of gravity, and hence move with a uniform velocity in a straight line. This theory could easily explain some of the observed facts such as reflection, refraction and rectilinear propagation of light.

Reflection. The phenomenon of reflection of light is explained as a result of the mechanical laws of perfectly *elastic impacts*, the incident corpuscles on striking the reflecting surface bounce off at angles equal to those at which they are incident. To explain the process of reflection Newton made an 'ad hoc' assumption that a reflecting surface exerts a force of repulsion on every panicle of light (corpuscle) incident on it in a direction normal to itself and this force decreases very rapidly with the increase of distance from the surface, so that it is effectively restricted only to a very thin layer near the surface bounded by the dotted line A_1, B_1.

Let PQ, be the path of a light corpuscle, having a uniform velocity V, incident upon the reflecting surface AB at an angle of incidence i. As the corpuscle approaches A_1B_1 that is, with in a very small distance from the surface AB, it begins to experience the repulsive force which continues to act, so long as it traverses this region of variable ether density, as it is generally called During its journey through this thin layer the component of the velocity of the particle parallel to the surface, *i.e.*, QE (=V sin i) remains unaffected while the normal component QN (—$V \cos i$) gradually diminishes, becomes zero and is finally reversed. It then gradually increases and regains its initial value when the particle reaches the boundary of this layer. Thus beyond R the particle moves away from the surface with the same uniform velocity V with which it was incident. *The path of the particle within this layer is curved.*

Since the component parallel to the surface remains unchanged during reflection, we have $QE=RF$

i.e., $V \sin i = V \sin r,$

or $\angle i = \angle r.$

Thus, *the angle of reflection is equal to the angle of incidence* and *by symmetry the incident ray, the reflected ray and the normal to the reflecting surface at the point of incidence are in the same plane.*

Refraction. The phenomenon of refraction of light was explained by Newton on the following assumptions :

(i) the corpuscles are attracted normally by the refracting surface AB of a denser medium when they come within a *very* thin layer bounded by A_1B_1 near the surface.

(*ii*) this attraction also continues to act within a *very* thin layer bounded by A_1B_2 inside the medium.

Accordingly the component of velocity normal to the surface goes on increasing during the passage of the corpuscle through the layer bounded between A_1B_1 and A_2B_2 till the corpuscle reaches the surface A_2B_2 within the medium where the attractive force ceases. After this it moves along RS with a uniform velocity, the parallel component of the velocity remaining unchanged. This explains the *bending* of the ray towards the normal in going from a rarer to a denser medium.

Suppose a light corpuscle is travelling along PQ. Let V_1 be its incident uniform velocity and i the angle of incidence in the rarer medium and V_2 its uniform velocity after refraction and r the angle of refraction in the denser medium. Since the components of velocities in the two media parallel to the refracting surface are the same, we have

$$QE = RF$$

i.e.,
$$V_1 \sin i = V_2 \sin r,$$

or
$$\frac{\sin i}{\sin r} = \frac{V^2}{V_1} = \mu \text{ (a constant)},$$

which is the Snell's law of refraction.

So if $\mu > 1$ as it is for glass or water, it means that $V_2 > V_1$. Thus according to the corpuscular theory the velocity of light should be greater in an optically denser medium, like glass and water, than in a rarer one, like air.

When the second medium is rarer than the first instead of the attractive force there is the repulsive force as assumed in explaining the phenomenon of reflection. But this repulsive force is not sufficient to make the normal component of velocity zero till after the corpuscle has entered the rarer medium and reached a layer where the repulsive force ceases and thereafter the corpuscle goes in a straight line but with a diminished velocity and bends away from the normal.

But actually the magnitudes of the velocities are just the reverse as proved by the experiments of Foucault on the determination of the velocity of light in water and carbon bi-sulphide. His results were in full agreement with the wave theory and in complete disagreement with the corpuscular theory. This theory also failed to explain other phenomena such as *interference, diffraction* and *polarization* of light. These shortcomings of the corpuscular theory were responsible for its downfall.

Newton's corpuscular theory, however, does not give any plausible explanation about the origin of repulsive and attractive forces normal to the surface while explaining the phenomena of reflection and refraction. In the case of transparent substances like glass, water, etc., light is partly reflected and partly refracted at the surface simultaneously. How it is possible for the surface to repel as well as to attract the particles of light at the same time ? To explain this difference in the behaviour of the corpuscles, Newton suggested the theory of fits, according to which when the corpuscles reach the surface, sometimes they are in a *fit of easy reflection,* and sometimes in *a fit of easy transmission* and pass periodically from one state to the other.

Newton also postulated the existence of a hypothetical medium, ether with variable density, being greater in free space and less in a material medium. According to him the velocity of the corpuscles depends upon the density of ether, being less in a rarer medium and greater in a denser one. Moreover, the impinging rays produce vibrations in the ether at the interfaces. These vibrations alternately contract and dilate the ether near the surface which respectively causes the reflection and refraction of the light corpuscles.

To explain the phenomenon of dispersion, Newton assumed that corpuscles of different colours have different sizes and that in a dispersive medium velocities of corpuscles depend upon their size. So that when white light passes through such a medium corpuscles associated with different colours deviate through different extents, thus bringing about dispersion.

Inspite of several arbitrary assumptions Newton's corpuscular theory could not explain the phenomena of interference, diffraction, polarization, double refraction and spectrometric data, etc.

Wave Theory

This theory was proposed by a Dutch citizen named Huygens, a contemporary of Newton, about the year 1678. According to him light is a form of wave-motion which is produced by mechanical vibrations of an all-pervading homogeneous medium called *luminiferous ether* just like those in solids and liquids. This artificial medium is supposed to be endowed with the properties of *elasticity* and *inertia* being thus able to transmit a wave motion. When these waves enter our eyes they excite the sensation of vision.

According to Huygens these waves were longitudinal and, as we would see later, he could quite successfully explain the phenomena of reflection, refraction, interference and diffraction Dispersion could also be interpreted if the velocity in a medium depended on the wavelength but Huygens could not account for the *approximate* rectilinear propagation of light and the phenomenon of polarization.

Fresnel and Young overcame this difficulty by assuming that light waves were transverse in nature. Fresnel explained the approximate rectilinear propagation of light by making use of secondary wavelets and dividing the wavefront into half-period zones, known as Fresnel's zones. To explain the transmission of transverse waves ether was supposed to possess the properties of an elastic solid. Since the velocity of a wave through an elastic medium should be equal to the square root of the elasticity divided by the density and for light this should equal 3×10^8 m/sec., it follows that ether must possess a very high elasticity and a very low density. It must be able to transmit the tiniest light impulse through vast interstellar spaces, yet it must in no way be disturbed by the earth whirling through it at a speed of about 30 kilometers per second. Obviously enough, fluids like water and air do not possess these properties. Thus there is no way out except assuming that ether is a solid. The velocities of transverse and longitudinal waves in a solid medium are given by

$$v_T = \sqrt{\frac{n}{d}}$$

and $$v_L = \sqrt{\frac{\left(k + \frac{4}{e} n\right)}{d}},$$

according as the strain is a shear or a uniform compression. Here n is the rigidity modulus, k the bulk modulus and d the density of the medium. There has been found no evidence of the existence of longitudinal disturbance in ether. To get over these difficulties ether is supposed to be highly "incompressible", *i.e.* $k = \infty$, so that, longitudinal vibration travelled with infinite velocity. On the other hand, Lord Kelvin suggested that $vL=0$, therefore, $k + \frac{4}{2} n = 0$, or $k = -\frac{4}{3} n$ *i.e.*, k is negative so that ether is "contractile". Thus, it is difficult to conceive a medium which can be stable under these conditions of contradictory properties which should be different in different media. However, the concept of elastic solid ether was very helpful in the theoretical development of the subject of light during the nineteenth century and it was with the advent of Maxwell's famous electromagnetic theory (1873) that the elastic solid theory was discarded. Till later in the last decade of the century certain other phenomena were discovered which demanded a quantum theory as proposed by Max. Planck, according to which light consists of discrete bundles of energy called *quanta* or *photons*.

For the purpose of our present discussion, we shall accept the wave-theory of light and begin with explanation of wavefronts and Huygen's principle of wave propagation.

Wavefronts and rays. It is very often convenient to represent a train of waves by means of wavefronts. Imagine a point source of light sends out waves in luminiferous' ether supposed to be homogeneous, isotropic and all -pervading medium in which matter exists. These waves will travel in all directions with the same velocity *c, i.e.,* the velocity of light and will reach a the particles around the source lying on the surface of a sphere drawn with the point as centre simulating easily. The surface of such a sphere is known as the wavefront and all the ether particles forming the wavefront to are in the same phase of vibration. Thus the waveforms, at any instant of time, is defined as the locus of all the particles in the me sum which are being disturbed at the same instant of time and are in the same phase of vibration. The intense cottons of a series of wavefronts emanating from a point source in the plane of the diagram. As the wave advances, the wavefronts move parallel to themselves in a direction perpendicular to them. At a great distance from the source the radii of curvature of the spherical wavefronts become large and small parts of the wavefronts can be treated as almost plane and we have a train of planewaves. The normal to the wavefront are known as rays. In the case of a train of plane waves the rays are all parallel to one another.

Huygens' Principle

Huygens' principle of wave propagation is a geometrical method of finding form the known shape and position of a wavefront at some instant, the shape and position it will have at some instant later. It is based on two postulates :

(*i*) *Each point on the wavefront is the centre of a new disturbance called secondary wavelet.* These wavelets travel out with the same velocity as the original wave os long as the medium remains the same.

(*ii*) *The envelope, viz., the tangential plane to these secondary wavelets constitutes the new wave front.*

Let us now explain the propagation of a wave on the basis of these postulates.

AB represents the trace of a spherical wavefront in the plane of the paper at any instant due to a point source of light S in a homogeneous and isotropic medium. It is required to find its position an instant *t* later. According to Huygens each point on the wavefront *AB* is the centre of a secondary wavelet. These wavelets advance with the same velocity in the homogeneous and isotropic ether as the original. In time *t*, after the disturbance has reached *AB*, each of these secondary wavelets will have moved a distance *ct*, where *c* is the velocity of propagation. To obtain the position of new wavefront describe a series of circles with centres *a*, *b*, *c*, etc., and radii *et*. The arc A_1B_1 in the plane of the paper enveloping these circles gives the trace of the new wavefront- All the points on the wavefront A_1B_1 *at any* instant, will be in the *same* phase of *vibration*. If the source of light is at a great distance, it will give rise to a plane wavefront *AB*. A new wavefront A_1B_1 can be drawn in the same manner as just explained.

It may be noted here that in accordance with Huygens' first postulate not only we also the *backward wavefront A'B'*. But Huygens further postulated *that the action of the secondary wavelets was confined to the points at ' which they touched the forward part of their envelope.* He thus considered only the forward wavefront and assumed that no backward wavefront, as such existed.

There was, however, no direct physical or mathematical justification for this arbitrary assumption to neglect all the unwanted parts of the secondary wavelets which tend to indicate that a plane wave front spreads out slightly at the edges giving rise to diffraction effects. The absence of a back-wave was explained many years later on Stokes' law according to which the amplitude propagated in any direction due to the secondary wave is proportional to $(1+\cos\theta)$, where θ is the angle between the wave normal and the direction of propagation. For the 'backward' wavefront $\theta = \pi$ and hence

$$(1+\cos\theta) = (1+\cos\pi) = (1-1) = 0$$

and so its effect is zero.

The waves of different wavelengths constituting a light beam, travel with the same velocity and so all the constituent waves coincide and move on as a single wave until the beam comes across a refracting medium through which the constituents travel with different velocities and undergo dispersion. This produces separate wavefronts which also travel with different velocities. Accepting Huygens' construction of the wave front we can satisfactorily explain the phenomena of reflection and refraction of light on the wave theory. But the explanation of rectilinear propagation of light on Huygens' principle of wave envelopes is not adequate, so a modified form of Huygens' principle as suggested by Fresnel is adopted.

Reflection of a plane wave at a plane surface. Suppose ADB, is the trace of a plane wavefront travelling towards a plane reflecting surface XY in the direction AA_1 and meeting it obliquely at A, E and B_1 then a, b, c normals to the wavefront are the rays of light. The incident wavefront AB and the reflecting surface XY are both perpendicular to the plane of the paper.

According to Huygens' Principle, when the wave reaches the surface, the secondary wavelets are immediately formed at the points touched so that every point of the surface XY in turn becomes the centre of secondary disturbance which is sent back into the medium. The wavefront first strikes the surface at A and lastly at the point B_1.

During the time the point B of the wavefront AB travels from B to B_1 (say t seconds) the secondary wavelet originating from A will have moved in the first medium a distance $Vt=AG = AA_1$ where V is the velocity of light in the first medium.

From B_1 draw a tangent plane to this wavelet. This will be the reflected wavefront if all the wavelets originating between A and B_1 touch this plane.

Now, in the absence of the surface XY, the wavefront AB would have occupied the position A_1B_1 after time t taken by the light to travel a distance equal to BB_1. The point D_x in A_1B_1 corresponds to the point D in AB. Thus ED_1 is the radius of the wavelet starting from E. To prove it draw EF perpendicular to B_1G, then in right-angled Δ 8 AGB_1 and $AA_1 B_1$ side $AA_1 = AG$, AB_1 is common to both. Hence Δ s are congruent and $\angle AB_1 A_1 = \angle AB_1 G$. Also the right-angled Δs EFB_1 and ED_1B_1 are congruent, for EB_1 is common to both and $\angle D_1B_1E = \angle FB_1E$ just proved.

Hence $EF = ED_1$, thus the disturbance from E touches the tangent plane B_1G and so is the case with all wavelets originating between A and B_1. Hence B_1G is the trace in the plane of the paper of the reflected wavefront.

Thus, a *plane wave is reflected as a plane wave from a plane surface.*

Since in right-angled Δs $AB_1 A_1$ AB_1G and AB_1B,

$$AA_1 = AG = BB_1 = Vt,$$

and AB_1 is common to all.

$\therefore$ the three triangles are equal in all respects.

Hence, $\angle BAB_1 = \angle GB_1A$

Since $\angle BAB_1$ is the angle between the incident wavefront and reflecting surface so it is equal to the angle between the incident ray and the normal, *i.e.*, it is equal to the angle of incidence (i). Similarly $\angle GB_1A$ is equal to the angle of reflection (r), that is

$$\angle i = \angle r.$$

Thus, the angle of incidence is equal to the angle of reflection.

Since AB, AB_1 and B_1G are all traces in the plane of paper of planes at right angles to the plane of the paper, the normal to them all lie in the *same* plane. As the normals to $A\ B$, B_1G and AB_1 are the incident ray, the reflected ray and the normal to the reflecting surface respectively, so it can be stated that the incident ray, the reflected ray and the normal to surface at the point of incidence all lie in the same plane. Thus the wave theory accounts for the laws of reflection.

Reflection of a spherical wave at a plane surface. Suppose XY, is the section of a plane reflecting surface at right angles to the plane of the paper, and ACB is the section of the incident spherical wave issuing from a point source of light O and touching the plane XY at C. By the time the waves from A reach A_1 and those from B reach B_1, the waves from C would have reached C_1 if the reflecting surface were not there. Now because of the presence of the reflecting surface the secondary wavelets originate from C and will have a radius equal to CD ($=CC_1$) and reach D. In turn each point in the wavefront ACB reaches the surface and sets up a disturbance. It is evident that these wavelets will touch the surface A_1DB_1 which becomes the position of the reflected wavefront with its centre of curvature at I which represents the *virtual* image of the point source O.

Since the wavefronts $A_1C_1\ B_1$ and A_1DB_1 are exactly similar in curvature to the wavefront ACB,

$$\therefore \qquad OC_1 = ID,$$

$$\text{or} \qquad OC + CC_1 = IC + CD,$$

$$i.e., \qquad OC = IC$$

$$(\text{since } CC_1 = CD).$$

Hence the *reflected image is as far behind the reflecting sur face as the object is in front of it.*

Sagitta theorem. If AMB, be a small arc of a circle of radius r and N is the mid-point of the chord AB, then MN is called the sagitta of the arc AMB. By geometry

$$AN^2 = MN \times NL$$

$$= MN\,(ML - MN)$$

$$= MN \times ML \text{ approx.,}$$

when MN is small.

$$\text{So that} \qquad MN = \frac{AN^2}{ML} = \frac{AN^2}{2r}.$$

As the curvature of an arc is reciprocal of the radius it follows that for spherical arcs on a common chord AB, for example, the sagitta MH is directly proportional to the curvature of the corresponding arc AMB.

Reflection of a spherical wave at a spherical surface. Suppose XPY, represents the section of the concave spherical reflecting surface with pole at P and centre of curvature at C. Let XAY be the section of an incident spherical wavefront issuing out from a point source of light O and touching the spherical surface at X and Y. The points X and Y become the sources of secondary wavelets which are sent back into the medium. By the time A wave at a spherical surface reaches P (say it takes t seconds) the secondary wavelets from X and Y reach D and E respectively so that $XD = YE = PA = Vt$. Similarly, the secondary wavelets from all other points in turn reach the corresponding points on the surface *DPE*, so that *DPE* is the section of the reflected wavefront with its centre of curvature at *I* which also gives the position of the *real* image of the point source *O*. Draw *DL* and *XM* perpendiculars to the principal axis *OP*.

Let *PO*, *PI* and *PC* be denoted by u, v and R respectively and *XM* and *DL* by y. Also since *XPY*, *XAY* and *DPE* are the arcs of circles of radii R, u and v respectively, therefore, by Sagitta theorem

$$PM = \frac{XM^2}{2PC} = \frac{y^2}{2R},$$

$$AM = \frac{XM}{2PO} = \frac{y^2}{2u}$$

and $$PL = \frac{DL^2}{2IP} = \frac{y^2}{2v}. \qquad [\therefore DL = XM]$$

Now, $$PL = PM + ML$$

$$= PM + XD = PM + PA$$

$$= PM + PM - AM = 2PM - AM$$

$$\therefore \quad PL + AM = 2PM.$$

Substituting the values of *PL*, *AM* and *PM* from above, we get conventionally.

$$-\frac{y^2}{2v} - \frac{y^2}{2u} = -\frac{2y^2}{2R}$$

or $$\frac{1}{v} + \frac{1}{u} = \frac{2}{R},$$

which is a relation between the distances of the object and image from the reflecting surface and its radius of curvature.

Refraction of a plane wave at a plane surface. Suppose *AB*, represents the section of a plane wavefront incident obliquely at an angle i on the plane surface of separation XY of the two media of refractive indices μ_1 and μ_2. Let V_1 and V_2 be the velocities of light in the two media respectively.

The wavefront ADB first of all touches the refracting surface at the point A and lastly at the point B_1. AB and XY represent only the traces of the planes at right angles to the plane of the paper.

According to Huygens' principle each point of the surface from A to B_1 in turn becomes the source of secondary wavelets as the wavefront AB sweeps over the surface XY. It is evident that in the absence of any change in the medium, the disturbance from A would have reached A_1 during the time the disturbance from B reached B_1 so that $AA_1=BB_1$ and A_1B_1 would have represented the trace of the wave front after a time t which the point B takes to reach B_1.

Now as soon as the wavefront reaches A the secondary wavelet starting from A, has travelled into the medium for t seconds when it is just going to start from B_1. To find the position of the refracted wavefront, with centre A and radius $V_2 t$ describe a sphere. Draw through B_1 a tangent plane to this sphere at right angles to the plane of the paper.

If B_1G is the trace of this plane in the plane of the diagram, we have to show that B_1G is the refracted wavefront, that is EF the perpendicular from E on B_1G is the radius of the wavelet from E when the surface is affected by the wavefront ADB.

Obviously, $AA_1 = V_1t$ and $AG = V_2t$.

$$\frac{AA_1}{AG} = \frac{V_1\,t}{V_2\,t} = \frac{V_1}{V_2} \qquad ...(i)$$

Again from similar Δs B_1AA_1 and $B_1\,EE_1$ and Δ s $B_1\,AG$ and $B_1\,EF_1$

$$\frac{EE_1}{AA_1} = \frac{B_1\,E}{B_1A} = \frac{EF}{AG}$$

$$\text{Consequently}\ \frac{EE_1}{EF} = \frac{AA_1}{AG} = \frac{V_1}{V_2}. \qquad \dots (ii)$$

$$\text{But}\ \frac{\textit{radius in medium } \mu_1}{\textit{radius in medium } \mu_2} = \frac{V_1}{V_2}.$$

and EE_1 is obviously the radius of the wavelet in the absence of the surface XY, *i e.,* in the medium μ_1 Thus EF is the radius of the wavelet in the medium μ_2 and as it is drawn perpendicular to B_1G the wavelet must touch this plane. Similarly, we can show that all the wavelets originating in AB_1 will touch the plane B_1G, which is therefore, the refracted wavefront.

It should follow from similar reasoning as in reflection that the incident ray and the refracted ray lie in the same plane with the normal to the surface.

Moreover, $\angle$s BAB_1 and GB_1A are the angles of incidence and refraction respectively since they are complements to the angles which the rays make with the surface, then

$$\frac{\sin i}{\sin r} = \frac{BB_1}{AB_1} \bigg/ \frac{AG}{AB_1}$$

$$= \frac{BB_1}{AG} = \frac{V_1}{V_2} = \text{constant.}$$

We thus find that the sines of the angles of incidence and refraction bear a constant ratio to each other which proves the Snell's law of refraction. This ratio is termed the refractive index of the second medium with respect to the first and is denoted by ${}^1\mu_2$.

$$\therefore \qquad {}^1\mu_2 = \frac{V_1}{V_2}$$

$$= \frac{\textit{Velocity of light in the first medium}}{\textit{Velocity of light in the second medium}}.$$

Further, when $r<i$, that is, the ray bends towards the normal during refraction, the ratio $\frac{\sin i}{\sin r}$ is greater than unity and the second medium is said to be more highly refracting than the first.

From the above equation it follows that the velocity of light in a highly refracting medium such as water is less than that in air or in vacuum, which is borne out by the experiments of Foucault and Michelson.

Refraction of a spherical wave on a plane surface. Let *APB*, represent the section of an incident spherical wave diverging from a luminous point *O*. When the wave reaches the boundary of separation *XPY* of the two media having refractive indices μ_1 and μ_2 respectively, then according to Huygens' construction secondary waves start in turn from all points as the wave passes over the surface. Had the wave continued in the same medium, then at some subsequent time *t* it would have occupied the position *XCY*, such that

$$PC = AX = BY.$$

But on account of the change in the medium as the point *A* moves to *X* and *B* to *Y* the point *P* moves to *D* only because the velocity of the wave inside the medium has changed from V_1 to V_2 Vi and V_2 respectively being the velocities of light in the two media. Thus the curvature of the wave front inside the medium has changed from *XCY* to *XDY so* that wave at a plane surface, the waves in the second medium appear to diverge from the point *I* which forms the image of the luminous point *O*.

Let *PO* and *PI* be represented by *u* and *v* respectively. Drop *AL* perpendicular on the principal axis and represent it by *y*.

Since *APB* and *XDY* form the arcs of circles of radii *OP* and *ID* respectively.

$$AL^2 = (2OP - PL)\ PL$$

$$= 2OP \times PL \ldots \text{approx.}$$

or $$XP^2 = 2OP \times PC,$$

$$\ldots. [\text{ since } AL = XP, \text{ and } PL = AX = PC$$

also $$XP^2 = (2ID - PD)\, PD,$$

or $$XP^2 = 2IP \times PD,$$

$$\ldots[PD \text{ is very small, so that } ID = IP]$$

$\therefore$ $$2OP \times PC = 2IP \times PD,$$

or $$u \times PC = v \times PD$$

or $$\frac{v}{u} = \frac{PC}{PD} - \frac{V_1 t}{V_2 t} = \left(\frac{\mu_2}{\mu_1}\right)$$

If $$\mu_1 = 1 \text{ and } \mu_2 = \mu, \text{ then}$$

$$\frac{v}{u} = \mu,$$

which is an important relation in geometrical optics.

Refraction of a spherical wave at a spherical surface. Suppose XAY, is the section of a spherical wave diverging from a luminous point O on the axis of a concave spherical surface XPY which separates the two media of refractive indices μ_1 (air) and μ_2 (glass). Let the wave touch the surface at X and Y.

If V_1 and V_2 respectively be the velocities of light in the two media, then

$$\frac{V_1}{V_2} = \frac{\mu_2}{\mu_1} = \mu. \qquad [\text{if } \mu_2 = \mu, \text{ and } \mu_1 = 1]$$

According to Huygens' construction when the wavefront touches the surface of separation, each point on the surface, in turn, becomes the source of secondary disturbance. The secondary wave lets from the points X and Y travel into the medium white the waves from other points on the wavefront first travel in air and then in the medium. By the time the disturbance from the point A teaches the pole P of the concave surface (where it enters the medium), the wavelets from X and Y reach B and D respectively into the medium so that BPD gives the section of the refracted wave front which appears to diverge from the point I which is the image of the luminous point O.

Since the time taken by the disturbance in going from A to P in air is the same as that taken by it from X to B or Y to D in the medium, so that

$$\frac{PA}{V_1} = \frac{XB}{V_2}$$

$$\therefore \quad BX = \frac{V_2}{V_1} PA = \frac{\mu_1}{\mu_2} PA \qquad ...(i)$$

Draw XL and BM, perpendiculars to the principal axis OP and represent XL and BM by y, PO by u, PI by v and PC by R.

Since XAY, XPY and BPD form arcs of circles with radii OA, CP and IP respectively, therefore, by Sagitta theorem

$$AL = \frac{XL^2}{2AO} = \frac{XL^2}{2PO} = \frac{y^2}{2u},$$

...[$\because AO = PO$ approx.

$$PL = \frac{XL^2}{2PC} = \frac{y^2}{2R}$$

and $$PM = \frac{BM^2}{2PI} = \frac{y^2}{2v}$$

Now, $$PM = PL - ML$$

$$= PL - XB \qquad ... [\because ML = XB]$$

$$= PL - \frac{\mu_1}{\mu_2} PA$$

...[by equation (i) above

or $\mu_2 PM = \mu_2 PL - \mu_1 PA$

$= \mu_2 PL - \mu_2 (PL - AL)$

... $[\because PA = PL - AL]$

$= (\mu_2 - \mu_1) PL + \mu_2 AL$

or $\mu_2 PM - \mu_1 AL = (\mu_2 - \mu_1) PL$

If $\mu_2 = 1$ and $\mu_2 = \mu$, then

$\mu_2 PM - AL = (\mu - 1) PL$

or $$\mu \frac{y^2}{2v} - \frac{y^2}{2u} = (\mu - 1) \frac{y^2}{2R}$$

or $$\frac{\mu}{v} - \frac{1}{u} = \frac{\mu - 1}{R}.$$

According to the sign conventions R, v and u are all *negative*.

$$\therefore \quad \frac{u}{-v} = \frac{1}{-u} = \frac{\mu - 1}{-R}$$

or $$\frac{\mu}{v} - \frac{1}{u} = \frac{\mu - 1}{R}.$$

If we proceed on the same lines, a similar relation is obtained for a convex spherical surface which is left as an exercise for students.

Total internal reflection on wave theory. The phenomena of total internal reflection can be satisfactorily explained on the wave theory. Let AB, be a plane wavefront in a denser medium (glass) incident at A separating it from air. Then as the wave-form sweeps over the interface the points between A and C are successively affected and become the origins of secondary wavelets in the upper and the lower media. The radius of the wavelet from A in the upper medium at the instant when B reaches C, is given by

$$AG = \frac{BC}{V^2} \times V_1 = \frac{V_1}{V_2} BC = \mu BC,$$

where V_1 and V_2 are the velocities of light in air and glass respectively and μ is the refractive index of glass.

But $$BC = AC \sin i,$$

where i is the angle of incidence,

$$\therefore \quad AG = \mu AC \sin i. \quad ...(i)$$

Now if we vary the angle of incidence, the following three cases arise :

(i) When $\mu \sin i < 1$, $AG < AC$ by equation (*i*) so that a tangent plane CG can be drawn from C to the secondary wavelets originating between A and C. CG, therefore gives the refracted wavefront.

(ii) When $\mu \sin i = 1$, $AG = AC$ by equation (*i*). This is the *limiting case* when a wavefront is formed. The secondary wavelet originating from A will pass through C at the instant when the latter is just affected and so is the case with all secondary wavelets originating between A and C. The refracted wavefront CD, therefore, is a plane perpendicular to the surface AC proceeding in a direction parallel to the surface of separation, the refracted ray grazing the surface. The angle i denned by $\sin i = 1/\mu$, is the critical angle for the more refracting medium.

(iii) When $\mu \sin i > 1$, that is, the angle of incidence is greater than the critical angle, $AG > AC$. No tangent plane can be drawn from the point C to any of the wavelets originating between A and C as the point C lies within them. Therefore, a refracted wavefront cannot exist and we get total internal reflection, the reflected wavefront CD proceeding in the direction of the arrow in the same medium.

Lens Formula on the Wave Theory

Convex lens. Suppose O is a luminous point on the principal axis of a convex lens, and XAY, the section of an incident spherical wavefront originating from O and just touching the lens surface at A. The secondary wavelets starting from the wave surface AT travel faster through air and slower through the denser medium like glass, so that during the time the secondary wavelets from A reach B through the lens, those from X and Y reach X_1 and Y_1 respectively. Thus $X_1 BY_1$ form the trace of the refracted wavefront which is assumed to be spherical and converges to the point I which forms the real image of the point O. Since the lens is *thin*, the paths XEX_1 and YFY_1 described in air are nearly straight an 1 parallel to the principal axis.

If V_1 and V_2 be the velocities of light in air and the material of the lens respectively and μ the refractive index of the lens material, then

$$\frac{V_1}{V_2} = \mu.$$

Since the time taken by the wavelets to travel through the material of the lens from A to B is the same as that taken by light from X to X_1 or Y to Y_1 in air, therefore, this time is given by

$$\frac{AB}{V^2} = \frac{XX_1}{V_1} = \frac{YY_1}{V_1}.$$

Draw XL, EM and X_1N perpendiculars to the principal axis of the lens, then $XX_1 = YY_1 = LN$.

$$\therefore \quad \frac{AB}{V^2} = \frac{LN}{V^1}$$

$$\text{or} \qquad LN = \frac{V_1}{V_2} AB = \mu\ AM$$

$$\text{or } LA + AB + 3N = \mu\ AB$$

or $\quad LA + BN = (\mu - 1)\, AB$

or $\quad LA + BN = (\mu - 1)\,(AM + MB)$ ...(*i*)

If R_1 be the radius of curvature of the first surface *EAF* at which the light enters the lens and R_2 that of the second surface *EBF* at which it emerges and *AO* and *BI* are represented by u and v respectively, then denoting XL ($=EM=X_1N$) by y and applying Sagitta theorem, we have

$$AL = \frac{y^2}{2u},\quad AM = \frac{y^2}{2R_1},$$

$$BM = \frac{y^2}{2R_2} \quad \text{and} \quad BN = \frac{y^2}{2v}$$

Substituting the values in equation (*i*) above, we have

$$\frac{y^2}{2u} + \frac{y^2}{2v} = (\mu - 1)\left(\frac{y^2}{2R_1} + \frac{y^2}{2R_2}\right)$$

or
$$\frac{1}{v} + \frac{1}{u} = (\mu - 1)\left(\frac{1}{R_1} + \frac{1}{R_2}\right).$$

According to the sign conventions used in the text, v and R_1 are *positive* and u and R_2 are *negative,* so that

$$\frac{1}{v} - \frac{1}{u} = (\mu - 1)\left(\frac{1}{R_1} + \frac{1}{R_2}\right).$$

Concave lens. Let a spherical wave *XAY*, diverging from a luminous point *O* on the principal axis of a thin concave lens touch the lens surface at *X* and *Y*. The concave lens being thinner at the centre than at the edges and also the velocity of light in the medium being less than that in air, during the time the disturbance travels from *X* to X_1 or *Y* to F_1 in the medium the wavelets starting from the point *A* will reach the point *B* travelling partly through air and partly through the medium. Thus $X_l BY_l$ is the trace of the refracted wavefront which is assumed to be spherical and appears to diverge from the point I which gives the position of the virtual image of the object point O.

Draw XL and $X_1 M$ perpendiculars to the principal axis of the lens. Let V_1 and V_2 be the velocities of the light in air and the medium respectively.

$$\therefore \qquad \frac{XX}{V^2} = \frac{AC}{V_1} + \frac{CD}{V_2} + \frac{DB}{V_1}$$

$$\text{Since} \qquad XX_1 = LM = LC + CD + DM$$

$$\therefore \qquad \frac{LC + CD + DM}{V_2} = \frac{AC}{V_1} + \frac{CD}{V_2} + \frac{DB}{V_1}$$

$$\text{or} \qquad \frac{LC}{V_2} + \frac{CD}{V_2} + \frac{DM}{V_2} = \frac{AC}{V_1} + \frac{CD}{V_2} + \frac{DB}{V_1}$$

$$\text{or} \qquad \frac{LC}{V_2} + \frac{DM}{V_2} = \frac{AC}{V_1} + \frac{DB}{V_1}$$

$$\text{or} \qquad = AC + DB = \frac{V_1}{V_2}(LC + DM),$$

$$\text{or} \qquad LC - LA + DM + MB = \mu (LC + DM),$$

$$\text{or} \qquad MB - LA = \mu (LC + DM) - (LC + DM)$$

$$\text{or} \qquad MB - LA = (\mu - 1)(LC + DM) \qquad \ldots (i)$$

If R_1 be the radius of curvature of the surface XCY at which the light enters the lens and R_2 that of the surface $X_1 DY_1$ at which it comes out, and AO and BI are denoted by u and v respectively, then denoting XL $(=X_1 M)$ by y, we have by sagitta theorem,

$$LC = \frac{y^2}{2R_1}, \quad DM = \frac{y^2}{2R_2},$$

$$LA = \frac{y^2}{2u} \text{ and } BM = \frac{y^2}{2v}.$$

Substituting the values in equation (i) above, we get

$$\frac{y^2}{2v} - \frac{y^2}{2u} = (\mu - 1)\left(\frac{y^2}{2R_1} + \frac{y^2}{2R_2}\right)$$

$$\frac{1}{v} - \frac{1}{u} = (\mu - 1)\left(\frac{1}{R_1} + \frac{1}{R_2}\right).$$

According to the sign conventions u, v and R_1 are negative while R_3 is positive, so that

$$\frac{1}{-v} - \frac{1}{-u} = (\mu - 1)\left(-\frac{1}{R_1} + \frac{1}{R_2}\right)$$

$$\frac{1}{v} - \frac{1}{u} = (\mu - 1)\left(\frac{1}{R_1} - \frac{1}{R_2}\right)$$

which is the same relation as deduced in the previous case.

Thus the formula

$$\frac{1}{v} - \frac{1}{u} = (\mu - 1)\left(\frac{1}{R_1} - \frac{1}{R_2}\right)$$

holds good both for a convex and a concave lens.

If the object lies at infinity, its image is formed at the focus of the lens, that is

when $$\frac{1}{u} = \frac{1}{\infty} = 0. \ \frac{1}{v} = \frac{1}{f}$$

Hence $$\frac{1}{f} = (\mu - 1)\left(\frac{1}{R_1} - \frac{1}{R_2}\right).$$

Example. *A convex lens of focal length 20 cm and made of glass of refractive index 1.5 is immersed in a liquid of refractive index 1.33. Find the change in the focal length of the lens.*

Here refractive index of glass, ${}^a\mu_g = 1.50$

Refractive index of liquid, $\quad {}^a\mu_l = 1.33.$

$\therefore$ refractive index of glass with respect to the liquid,

$$ {}^l\mu_g = {}^l\mu_i \times {}^a\mu_g = \frac{{}^a\mu_g}{{}^a\mu_1} = \frac{1.50}{1.33} = 1.128. $$

Focal length of the convex lens in air, $f = 20$ cm.

If F denotes the focal length of the lens in liquid and R_1 and R_2 the radii of curvature of its surfaces, then

$$ \frac{1}{f} = (1.50 - 1)\left(\frac{1}{R_1} - \frac{1}{R_2}\right) $$

and
$$ \frac{1}{F} = (1.128 - 1)\left(\frac{1}{R_1} - \frac{1}{R_2}\right) $$

$\therefore$
$$ \frac{F}{f} = \frac{1.50 - 1}{1.128 - 1} = \frac{0.50}{0.128}, $$

or
$$ F = \frac{0.50}{0.128} \times 20 \text{ cm} $$

$$ = 78.12 \text{ cm.} \qquad \text{... } [\because\ f = 20 \text{ cm}] $$

Thus the change in focal length=(78.12 – 20.0)=58 – 12 cm.

Note. For more examples the student is advised to refer to part [(Geometrical Optics) of the book.

8

Refraction of Light

Refraction of light. When a ray of light passes from one transparent medium to another, in which it has a different velocity, there occurs a change in the direction of propagation of light except when it strikes the surface of separation of two media normally. This bending of a ray of light is called refraction.

The angles made by the incident ray and the refracted ray with the normal to the separating surface at the point of incidence are called the angles, of incidence and of refraction respectively. The refraction takes place according to the following laws :

Law 1. The incident ray, the normal and the refracted ray all lie in the same plane.

Law 2. The ratio of the sine of angle of incidence to the sine of angle of refraction for any two media is constant for a light of definite colour.

The second law was discovered by Willebrod Snellius in 1621 and is known after his name as Snell's Law.

Denoting the angles of incidence and refraction by i and r respectively, the second law states that

$$\frac{\sin i}{\sin r} = {}^1\mu_2,$$

where μ is a constant depending upon the nature of media and the kind of light. It is called the refractive index of the second medium with respect to the first, the subscripts 1 and 2 indicating that the light passes from medium 1 to medium 2. If the direction of light is reversed, the ray will retrace its path, so that

$$\frac{\sin r}{\sin i} = {}^2\mu_1,$$

$$\therefore \qquad {}^1\mu_2 \times {}^2\mu_1 = 1 \qquad \text{... (1)}$$

When a ray enters from vacuum into a medium, the above ratio is called the *absolute refractive index* of that medium.

The term refractive index also conveys a *physical idea*, as on the wave theory of light it expresses the ratio of the velocity of light in vacuum to its velocity in the medium, *viz.*,

$$\mu = \frac{\text{velocity of light in vacuum}}{\text{velocity of light in medium}} \qquad \text{... (2)}$$

Similarly, the refractive index of a medium B with respect to a medium A may be written as

$${}^A\mu_B = \frac{\text{velocity of light in medium } A}{\text{velocity of light in medium } B} \qquad \text{...(3)}$$

The refractive index of a medium varies with wavelength, optical density and temperature etc.

Snell's Law

Snell's law is sometimes expressed in a more symmetrical form deduced as follows :

Let μ_1 and μ_2 be the refractive indices for two media with respect to air, then

$$\mu_1 = \frac{\text{velocity of light in air}}{\text{velocity of light in medium (1)}} = \frac{C}{C_1}$$

and
$$\mu_2 = \frac{\text{velocity of light in air}}{\text{velocity of light in medium (2)}} = \frac{C}{C_2}$$

where C is the velocity of light in air, C_1 and C_2 the respective velocities of light in the two media.

Dividing, we have

$$\frac{\mu_2}{\mu_1} = \frac{C_1}{C_2} = {}^1\mu_2$$

[by definition of refractive index]

By Snell's law

$${}^1\mu_2 = \frac{\sin i}{\sin r},$$

$$\therefore \qquad \frac{\mu_2}{\mu_1} = \frac{\sin i}{\sin r},$$

or $$\mu_1 \sin i = \mu_2 \sin r.$$

That is, the product of refractive index and the sine of the angle made by the ray with normal at the point of incidence is constant for a given ray in both media.

Fermat's Principle

Optical path and Fermat's principle of least time. When a ray of light travels a distance S in a medium of refractive index μ, the product μS is said to be the optical path or the equivalent air path Since the value of μ for air does not appreciably differ from that for vacuum, it represents the distance in vacuum that light would travel in the same time in which it travels the distance S in the medium. .

When a ray travels distances S_1, S_2, S_3 etc. in media of indices μ_1, μ_2, μ_3 etc., then the optical path is given by

$$S = \mu_1 S_1 + \mu_2 S_2 + \mu_3 S_3 + \ldots\ldots$$
$$= \Sigma \mu_1 S_1.$$

For a medium of continuously varying optical density, the optical path of a ray PQ is expressed as an integral,

$$S = \int_P^O \mu dS. \quad \ldots(5)$$

Fermat's principle. In its original form it was stated as follows :

When a ray of light goes from one point to another through a set of media, it always follows that path along which the time taken is minimum.

It should be noted that this original statement of the principle by Fermat is not complete and has been modified, for there are a number of cases where the optical path is *maximum* or else neither a maximum nor a minimum but *stationary* as in the case of formation of images by a lens in which all rays proceeding from an object point traverse the same *constant* path to the corresponding image point.

For a homogeneous and isotropic medium, μ being constant, we have

$$\int_P^Q dS = \text{a minimum, } i.e., \delta \int_P^Q dS = 0.$$

which means that between two points a ray will travel in a straight line and not any other path. Thus in such a medium *rectilinear* propagation of light is explained.

For a non-homogeneous and anisotropic medium

$$\int_P^Q \mu dS = \text{a minimum or a maximum or stationary,}$$

$$i.e., \text{ again } \delta \int_P^Q \mu dS = 0. \qquad \ldots (6)$$

This is the analytical expression of the second way of stating the law that *"for refraction through prisms and lenses the path of the ray between two points is that for which the time is* stationary *in value".*

Derivation of Laws of Reflection and Refraction from Format's Principle

Law of reflection. Let PA be the incident ray and AQ the reflected ray ; i and r the angles of incidence and reflection respectively. We have to show that the law of reflection, *viz.,*

$$\angle i = \angle r$$

is compatible with the principle of least time.

Let the perpendicular distances PL and QM be denoted by h_1 and h_2 respectively and the total length on X-axis intercepted by these perpendiculars, *i.e., LM by p.*

The optical path PAQ is given by

$$S = PA + AQ$$

Taking $\qquad AM = x$

and $\qquad AL = (p - x)$, we get

$$S = \left[h_1^2 + (p-x)^2\right]^{\frac{1}{2}} + \left[h_2^2 + x^2\right]^{\frac{1}{2}}$$

Differentiating and equating the first derivative to zero, we have

$$\frac{dS}{dx} = \frac{1}{2}\frac{2(p-x)\times(-1)}{\left[h_1^2 + (p-x)^2\right]^{\frac{1}{2}}} + \frac{1}{2}\frac{2x}{\left(h_2^2 + x^2\right)^{\frac{1}{2}}} = 0,$$

$$\therefore \frac{p-x}{\left[h_1^2+(p-x)^2\right]^{\frac{1}{2}}} = \frac{x}{\left(h^2+x^2\right)^{\frac{1}{2}}}$$

i.e., $\sin i = \sin r$

or $i = r.$

Law of refraction. PAQ is the path of a ray from the medium of refractive index μ_1 into a medium of refractive index μ_2 passing *via* a point A at the interface, then the optical path, PAQ is given by

$$S = \mu_1 PA + \mu_2 AQ.$$

Taking A as origin, $AM = x$ and $AL = (p - x)$, then

$$S = \mu_1\left[h_1^2+(p-x)^2\right]^{\frac{1}{2}} + \mu_2\left[h_2^2+x^2\right]^{\frac{1}{2}}$$

According to Fermat's principle S is either a maximum or a minimum or stationary.

In either case the first derivative of the expression above must vanish, that is

$$\frac{dS}{dx} = 0$$

Now

$$\frac{dS}{dx} = \frac{\frac{1}{2}\mu_1}{\left[h_1^2+(p-x)^2\right]^{\frac{1}{2}}}\times(-2p+2x)$$

$$+\frac{\frac{1}{2}\mu_2}{\left(h_2^2+x^2\right)^{\frac{1}{2}}}\times 2x = 0,$$

$$\therefore \mu_1\frac{p-x}{\left[h_1^2+(p-x)^2\right]^{\frac{1}{2}}} = \frac{\mu_2 x}{\left(h_2^2+x^2\right)^{\frac{1}{2}}}$$

Since $$\frac{p-x}{\left[h_1^2+(p-x)^2\right]^{\frac{1}{2}}} = \sin i$$

and $$\frac{x}{\left[h_2^2+x^2\right]^{\frac{1}{2}}} = \sin r.$$

$$\therefore \qquad \mu_1 \sin i = \mu_2 \sin r$$

The Snell's law is, therefore, proved.

Aplanatic surface. An aplanatic surface has been defined, as one in which all rays originating from a point source P are brought to focus after reflection at the surface to some other point P'. These points P and P' are called aplanatic foci. These definitions hold for refracting surfaces as well. According to fermat's principle, the optical paths of all these rays must be equal so that if the points P and P' lie in different media the equation for the aplanatic surface becomes

$$\mu_1 PA + \mu_2 AP' = \text{constant} \qquad \text{... (7)}$$

We can easily prove it for a single refraction. Let CD be an interface between two media of refractive indices μ_1 and μ_2 and PAP' the path of a ray from P to P' meeting the interface at A as we have to show that

$$\mu_1 PA + \mu_2 AP'$$

is stationary for the actual path. Let B be any point on CD close to A and let α_1 and α_2 be the complements of the angles PAB and $P'BA$, then obviously, α_1 the angle of incidence of the ray PA and α_2 the angle of refraction of the ray PB. But in the limit when AB is made infinitely small, α_2 is also the angle of refraction of the ray PA so that

$$\mu_1 \sin \alpha_1 = \mu_2 \sin \alpha_2$$

by Snell's law. Draw BL and AM perpendiculars on PA and BP'.

$$\begin{aligned}\text{Now } (\mu_1\, PA + \mu_2\, AP') - (\mu_1\, PB + \mu_2\, BP') &= \mu_1\,(PA - PB) + \mu_2\,(AP' - BP') \\ &= \mu_1\, AL - \mu_2 BM \\ &= \mu_1\, AB \cos \angle LAB - \mu_2 AB \cos \angle ABM \\ &= \mu_1\, AB \sin \alpha_1 - \mu_2\, AB \sin \alpha_2 \\ &= AB\left[\mu_1 \sin \alpha_1 - \mu_2 \sin \alpha_2\right] \\ &= 0, \text{ since } \mu_1 \sin \alpha_1 = \mu_2 \sin \alpha_2\end{aligned}$$

as stated above.

$$\mu_1 PA + \mu_2\, AP' = \mu_1 PB + \mu_2 BP'$$

Hence the expression $\mu_1 PA + \mu_2\, AP'$ is constant for small displacements of the point A and the locus of A is called an aplanatic surface. Such surfaces were first investigated by Descartes and are called *Cartesian ovals.*

In the case of reflection, the condition for the aplanatic surface becomes

$$PA + AP' = \text{constant, since } \mu_1 = \mu_2 \qquad \text{... (8)}$$

The properties of such a surface were briefly referred. For further details regarding aplanatic surfaces.

The application of Fermat's principle to refraction at a single surface and refraction through a lens is dealt with.

Refraction through spherical surfaces. When a widely divergent pencil of rays is incident on a spherical refracting surface, separating two media, the refracted rays do not generally intersect in a single point but they touch a *caustic surface*, somewhat similar to that which occurs when reflection takes place at a spherical surface (Art. 1-6). If, however, a point be taken on the principal axis of the refracting surface and if we confine our attention to a very small centric pencil of rays incident on the surface, it will be found that after refraction the pencil will either converge to or appear to diverge from a single point.

Let us deduce relations connecting object and image distances in such cases. Here, we shall use the same general methods and conventions etc. as employed in reflection from spherical surfaces.

Light incident on a convex spherical surface. Let AMB represent a convex spherical surface of radius of curvature r separating a less dense medium of refractive index μ_1 on the left from the more dense medium (shaded) of refractive index μ_2 on the right so that $\mu_2 > \mu_1$. Let C be the centre of curvature of the surface. Let P be a point source on the principal axis and PA a ray incident at A which after refraction, meets or appears to meet the axis in Q. θ and ϕ are the angles of incidence and refraction respectively and α, β, γ are the angles which the incident ray, the refracted ray and the normal at A make with the principal axis. Then, by Snell's law,

$$\mu_1 \sin\theta = \mu_2 \sin\phi,$$

or if the angles are small

$$\mu_1\theta = \mu_2\phi. \qquad \text{... } (a)$$

In ΔAPC, θ is the exterior angle, so that

$$\theta = \alpha + \gamma.$$

Similarly, for ΔAQC

$$\gamma = \phi + \beta,$$

or

$$\phi = \gamma - \beta.$$

Substituting the values of θ and ϕ in (a), we have

$$\mu_1 (\alpha + \gamma) = \mu_2 (\gamma - \beta),$$

Transposing, we get

$$\mu_2\beta + \mu_1\alpha = (\mu_2 - \mu_1)\,\gamma.$$

Denoting *MA* by *y* and applying the coordinate geometry conventions,

$$\alpha = \frac{y}{-u};$$

$$\beta = \frac{y}{v};$$

$$\gamma = \frac{y}{r}, \qquad [\because u \text{ is } -\text{ve}, v \text{ and } r \text{ are } +\text{ve}]$$

$\therefore$ from equation (*b*) $\mu_2 \dfrac{y}{v} + \mu_1 \dfrac{y}{-u} = (\mu_2 - \mu_1)\dfrac{y}{r}$,

$$\frac{\mu_2}{v} - \frac{\mu_1}{u} = \frac{\mu_2 - \mu_1}{r} \qquad \text{... (9)}$$

If the first medium be air, the formula reduces to

$$\frac{\mu}{v} - \frac{1}{u} = \frac{\mu - 1}{r}. \qquad \text{... (9a)}$$

We have

$$\theta = \alpha + \gamma$$

and

$$\phi = \beta + \gamma,$$

$$\therefore \quad \mu_1(\alpha + \gamma) = \mu_2(\beta + \gamma),$$

or

$$-\mu_2\beta + \mu_1\alpha = (\mu_2 - \mu_1)\gamma.$$

Substituting the values of α, β and γ with due regard being paid to conventional signs, we have

$$-\mu_2 \frac{y}{-v} + \mu_1 \frac{y}{-u} = (\mu_2 - \mu_1)\frac{y}{r}$$

$$\frac{\mu_2}{v} - \frac{\mu_1}{u} = \frac{\mu_2 - \mu_1}{r},$$

the same relation as (9) above.

Light incident on a concave spherical surface. Here the incident ray *PA* after refraction appears to intersect the axis in *Q*. With the same notation, we have as before,

$$\mu_1\theta = \mu_2\phi \qquad \text{... }(a)$$

from above

$$\theta = \gamma - \alpha$$

and
$$\phi = \gamma - \beta$$

Putting the values of θ and ϕ in (*a*) above, we have

$$\mu_1 (\gamma - \alpha) = \mu_2(\gamma - \beta).$$

Transposing

$$\mu_2\beta - \mu_1\alpha = (\mu_2 - \mu_1)\, \gamma. \qquad \text{... }(b)$$

Now,
$$\alpha = \frac{y}{-u},$$

$$\beta = \frac{y}{-v},$$

$$\gamma = \frac{y}{-r}$$

$\because$ *u*, *v*, *r* are all negative being on the left of origin,

$\therefore$ from equation (*b*)

$$\frac{\mu_2 y}{-v} - \frac{\mu_1 y}{-u} = (\mu_2 - \mu_1)\frac{y}{-r}$$

or on simplifying,

$$\frac{\mu_2}{v} - \frac{\mu_1}{u} = \frac{\mu_2 - \mu_1}{r},$$

or if the first medium be air, the formula becomes

$$\frac{\mu}{v} - \frac{1}{u} = \frac{\mu - 1}{r},$$

relations similar to (9) and (9*a*) above for a convex spherical surface.

For *large angles* of incidence and refraction and assuming the object and image distances u and v respectively measured along the rays, the correct formula would become,

$$\frac{\mu_2}{v} - \frac{\mu_1}{u} = \frac{\mu_2 \cos \phi - \mu_1 \cos \theta}{r}.$$

It may be deduced for a concave surface as below :

According to our conventions

$$AP = -u;$$

$$AQ = -v$$

and $$AC = -r.$$

Now $$\Delta PAC = \Delta PAQ + \Delta QAC,$$

or trigonometrically,

$$\frac{1}{2} AP.AC \sin \theta = \frac{1}{2} AP.AQ \sin (\theta - \phi) + \frac{1}{2} AQ.AC \sin \phi.$$

or $$ur \sin \theta = uv \sin (\theta - \phi) + vr \sin \phi \qquad \text{... } (c)$$

Also, $$\frac{\sin \theta}{\sin \phi} = \frac{\mu_2}{\mu_1} \qquad \text{[Snell's law]}$$

Expanding Eqn. (*c*), we have

$$ur \sin \phi = uv \sin \theta \cos \phi$$

$$- uv \cos \theta \sin \phi + vr \sin \phi.$$

Dividing both sides by $uvr \sin \phi$ and rearranging, we have

$$\frac{1}{v} \frac{\sin \theta}{\sin \phi} = \frac{1}{r} \frac{\sin \theta}{\sin \phi} \cos \phi - \frac{1}{r} \cos \theta + \frac{1}{u}$$

Substituting $\frac{\mu_2}{\mu_1}$ for $\frac{\sin\theta}{\sin\phi}$, we get

$$\frac{1}{v}\frac{\mu_2}{\mu_1} = \frac{\cos\phi}{r}\cdot\frac{\mu_2}{\mu_1} - \frac{1}{r}\cos\theta + \frac{1}{u},$$

or
$$\frac{\mu_2}{v} = \frac{\cos\phi}{r}\cdot\mu_2 - \frac{\mu_1\cos\theta}{r} + \frac{\mu_1}{u},$$

or
$$\frac{\mu_2}{v} - \frac{\mu_1}{u} = \frac{\mu_2\cos\phi - \mu_1\cos\theta}{r} \qquad \text{... (10)}$$

if the first medium be air, we have

$$\frac{\mu}{v} - \frac{1}{u} = \frac{\mu\cos\phi - \cos\theta}{r} \qquad \text{... (10}a\text{)}$$

The derivation of a similar result for a convex surface is left as an exercise for the reader.

Conjugate foci relation for refraction at a spherical surface by Fermat's principle. Let *C*, be the centre of curvature of a convex spherical surface separating two isotropic media of refractive indices μ_1 and μ_2 respectively. Take *M* as the origin according to the coordinate geometry conventions adopted in the text. The equation to the section of the separating spherical surface in the plane of the diagram is

$$(x - r)^2 + y^2 = r^2$$

or
$$x^2 + y^2 = 2rx,$$

where $(r, 0)$ are the coordinates of *C*.

Let $P\ (u, 0)$ be the point object on the axis and $(v, 0)$ its image formed by refraction. Let *PA* be an incident ray at a point $A\ (x, y)$ on the surface and *AQ* the corresponding refracted ray. According to Fermat's law, the time taken from *P* to *Q* has an extreme value.

Now,

$$AP = \left[(x-u)^2 + y^2\right]^{\frac{1}{2}}$$

... [– ve sign being due to our convention

$$= -\left[x^2 + u^2 - 2xu + y^2\right]^{\frac{1}{2}}$$

$$= -\left[u^2 - 2xu + 2rx\right]^{\frac{1}{2}}$$

by the equation given above

$$= -u\left[1 - \left(\frac{2x}{u} - \frac{2rx}{u^2}\right)\right]^{\frac{1}{2}}$$

$$= -u\left[1 - \frac{x}{u} + \frac{rx}{u^2}\right]$$

$$= -\left[u - x + \frac{rx}{u}\right]$$

by binomial theorem, since all the terms except the first are small if paraxial rays alone are considered.

Similarly, $$AQ = \left[(v-x)^2 + y^2\right]^{\frac{1}{2}}$$

$$= v\left[1 - \frac{x}{v} + \frac{rx}{v^2}\right]$$

$$= \left[v - x + \frac{rx}{v}\right]$$

Now $\mu_1\, AP + \mu_2\, AQ = \text{constant}$, by Fermat's principle. Then inserting the values of AP and AQ just obtained, we have

$$-\mu_1\left[u - x + \frac{rx}{u}\right] + \mu_2\left[v - x + \frac{rx}{v}\right] = \text{constant.}$$

Differentiating with respect to x, we get

$$-\mu_1\left[-1+\frac{r}{u}\right]+\mu_2\left[-1+\frac{r}{v}\right] = 0,$$

or
$$\mu_1 - \frac{\mu_1 r}{u} - \mu_2 + \frac{\mu_2 r}{v} = 0,$$

or
$$\frac{\mu_2 r}{v} - \frac{\mu_1 r}{u} = \mu_2 - \mu_1,$$

whence
$$\frac{\mu_2}{v} - \frac{\mu_1}{u} = \frac{\mu_2 - \mu_1}{r},$$

or if the first medium be air, we have

$$\frac{\mu}{v} - \frac{1}{u} = \frac{\mu - 1}{r},$$

which are the usual refraction relations between conjugate foci.

By setting μ_1 for $-\mu_2$ and -1 for μ in the above obtain the relation we obtain the relation

$$\frac{1}{v} + \frac{1}{u} = \frac{2}{r},$$

giving the positions of conjugate foci in the case of reflection at a spherical surface.

Proceeding in a similar way we can derive these relations for a concave refracting surface. The reader is expected to attempt these solutions himself.

Power of a Surface

The expression $\frac{(\mu_2 - \mu_1)}{r}$, is called the power of a refracting surface of refractive index μ_2 lying in a medium of refractive index μ_1. According to our convention the power of a convex surface is positive, since both $(\mu_1 - \mu_2)$ and r are positive while for a concave surface the power is negative because $(\mu_1 - \mu_2)$ is positive and r is negative. In general the power of a refracting surface which converges the incident parallel light is positive and that of a surface which diverges it, is negative. Denoting it by P, we have from equation (9).

$$P = \frac{\mu_2 - \mu_1}{r}$$

$$= \frac{\mu_2}{v} - \frac{\mu_1}{u}, \qquad \text{... (11)}$$

or in case the first medium is air,

$$P = \frac{\mu - 1}{r} = \frac{\mu}{v} - \frac{1}{u}. \qquad \text{... (11a)}$$

If r, u and v are expressed in meters, the power is expressed in diopters.

Principal foci and principal focal planes. Equation (9) gives us the position of two important points on the principal axis called the first and the second principal foci F_1 and F_2 of the surface respectively. F_1 is the position of the object for which the image is at infinity. If we denote its distance from M by f_1, then by making v infinitely great, we have

$$\frac{\mu_2}{\infty} - \frac{\mu_1}{f_1} = \frac{\mu_2 - \mu_1}{r}$$

or

$$\frac{\mu_1}{f_1} = -\frac{\mu_2 - \mu_1}{r}$$

$$\therefore \quad f_1 = -\frac{\mu_1 r}{\mu_2 - \mu_1} \qquad \text{... (12)}$$

If the first medium is air $\mu_1 = 1$, and $\mu_2 = \mu$ say, then

$$f_1 = \frac{r}{\mu - 1}. \qquad \text{... (12a)}$$

Similarly, if the object is at infinity the position of the image is termed the second principal focus F_2. If f_2 be its distance from M, then

$$\frac{\mu_2}{f_2} - \frac{\mu_1}{\infty} = \frac{\mu_2 - \mu_1}{r},$$

or
$$\frac{\mu_2}{f_2} = \frac{\mu_2 - \mu_1}{r}$$

$\therefore$
$$f_2 = \frac{\mu_2 r}{\mu_2 - \mu_1} \qquad \text{... (13)}$$

If the first medium be air, then

$$f_2 = \frac{\mu r}{\mu - 1} \qquad \text{... (13}a\text{)}$$

An incident ray proceeding from the first principal focus of the surface gives rise to a refracted ray parallel to the axis and a ray parallel to the axis gives rise to a refracted ray proceeding towards the second principal focus of the surface.

Planes through F_1 and F_2, at right angles to the principal axis, are called the first and the second principal focal planes of the refracting surface.

From equations (12) and (13), we also get

$$\frac{f_1}{f_2} = -\frac{\mu_1}{\mu_2}$$

or
$$\frac{f_1}{\mu_1} + \frac{f_2}{\mu_2} = 0,$$

or
$$\mu_1 f_2 + \mu_2 f_1 = 0 \qquad \text{... (14)}$$

If first medium is air

$$\mu_1 = 1$$

and
$$\mu_2 = \mu.$$

$$\therefore \qquad f_2 + \mu f_1 = 0 \qquad \text{... (14a)}$$

The distinction between the two focal distances should be carefully noted.

From (12*a*) $\qquad -(\mu - 1) f_1 = r$

or numerically $\qquad (\mu - 1) f_1 = r$

and from (14a) $\qquad f_2 = -\mu f_1,$

or numerically $\qquad f_2 - f_1 = \mu f_1 - f_1 = (\mu - 1) f_1$

That is, the numerical difference between the first and second focal distances is equal to the numerical value of the radius of curvature.

Also equation (9) can be expressed in the following forms:

(*i*)
$$\frac{\mu_2}{v} - \frac{\mu_1}{u} = \frac{\mu_2 - \mu_1}{r} = -\frac{\mu_1}{f_1}, \left[\because f_1 = -\frac{\mu_1 r}{\mu_2 - \mu_1}\right]$$

$$\frac{\mu_2}{v} - \frac{\mu_1}{u} = \frac{\mu_2 - \mu_1}{r} = \frac{\mu_2}{f_2}. \left[\because f_2 = -\frac{\mu_2 r}{\mu_2 - \mu_1}\right]$$

Either of these equations gives the conjugate distances for a single spherical surface and is analogous to the Gaussian formula for a thin lens.

(*iii*) Multiplying equation (9) by $\frac{r}{\mu_2 - \mu_1}$, we have

$$\frac{\mu_2 r}{(\mu_2 - \mu_1)v} - \frac{\mu_1 r}{(\mu_2 - \mu_1)u} = 1,$$

whence by equations (12) and (13)

$$\frac{f_2}{v} + \frac{f_1}{u} = 1, \qquad \text{... (15)}$$

or $\quad uv - vf_1 - uf_2 = 0$

and on adding $f_1 f_2$ to both sides of this equation, we get

$$(u - f_1)(v - f_2) = f_1 f_2, \qquad \text{... (15a)}$$

a relation connecting u, v, f_1, and f_2.

Image of a small object by refraction at a spherical surface. Consider a small object pP_1, in front of a convex spherical surface AMB. Let C be its centre of curvature, F_1 and F_2, the positions of the first and second principal foci. A ray parallel to the axis after refraction passes through the second principal focus F_2. The ray P_1C directed towards the centre of curvature goes undeviated and meets the other ray in Q_1 which is, therefore, the image of the point P_1. All other rays from P_1 will also be brought to focus at Q_1. For instance, the ray P_1F_1 which passes through the first principal focus will by definition of F_1 be refracted parallel to the axis and meet other rays in Q_1. The rest of the image lies in the conjugate plane through this point.

Transverse or lateral magnification. It is the ratio of any transverse dimension of the image to the corresponding transverse dimension of the object.

From right-angled triangles CPP_1 and CQQ_1

$$\frac{QQ_1}{PP_1} = \frac{CQ}{CP}$$

$$= \frac{MQ - MC}{MP + MC}$$

or
$$-\frac{h_2}{h_1} = \frac{v - r}{-u + r},$$

where $QQ_1 = MB = -h_2$

and $PP_1 = AM = h_1$

But $\frac{h_2}{h_1}$ is the *transverse magnification* denoted by m,

$$\therefore \qquad m = \frac{h_2}{h_1} = -\frac{v - r}{-u + r} = \frac{v - r}{u - r} \qquad \text{... (16)}$$

If m is positive, the image is virtual and erect, while if m is negative, the image is real and inverted.

Another useful expression for m is derived by making u se of the ray P_1M, which on refraction also meets at Q_1.

Such that $\angle P_1MP = \theta$ = angle of incidence

and $\angle Q_1MQ = \phi$ = angle of refraction.

Now $$\frac{QQ_1}{PP_1} = \frac{\text{MQ} \tan \phi}{\text{MP} \tan \theta}$$

or, *conventionally*

$$-\frac{h_2}{h_1} = \frac{v \tan \phi}{-u \tan \theta}.$$

$$\therefore \quad m = \frac{h_2}{h_1} = \frac{v}{u} \cdot \frac{\tan \phi}{\tan \theta}$$

$$= \frac{v}{u} \cdot \frac{\sin \phi}{\sin \theta}$$

$$= \frac{v}{u} \times \frac{\sin \phi}{\sin \theta} \qquad [\because \theta \text{ and } \phi \text{ are small}]$$

$$= \frac{v}{u} \frac{\mu_1}{\mu_2} \qquad \text{... [by Snell's law]}$$

$$= \frac{\text{v}}{\mu_2} \Big/ \frac{\text{u}}{\mu_1} \qquad \text{... (16a)}$$

where v/μ_2 and u/μ_1 are the reduced image and object distances in respective media.

If first medium is air, then $\frac{\mu_2}{\mu_1} = \mu$,

$$\therefore \quad m = \frac{v}{u\mu}. \qquad \text{... (16b)}$$

Axial Magnification

It is the ratio of the depth of the image along the axis to the corresponding depth of the object.

Denoting it by m', we have

where u and v are the object and image distances from the surface.

Differentiating the image object relation

$$\frac{\mu_2}{v} - \frac{\mu_1}{u} = \frac{\mu_2 - \mu_1}{r}, \text{ we get}$$

$$\mu_2 \frac{dv}{v^2} - \mu_1 \frac{du}{u^2} = 0,$$

$$\text{and, therefore,} m' = \frac{dv}{du} = \frac{\mu_1}{\mu_2} \cdot \frac{v^2}{u^2}$$

$$= \frac{\mu_2}{\mu_1}\left(\frac{\mu_1 v}{\mu_2 u}\right)^2 = \frac{\mu_2}{\mu_1} m^2, \qquad \text{... (17)}$$

where m is the transverse magnification. As $\mu_1 = 1$ for air then

$$m' = \mu m^2. \qquad \text{... (17a)}$$

Also by differentiation of equation (15), we have

$$m' = \frac{dv}{du} = -\frac{v^2}{u^2} \cdot \frac{f_1}{f_2}. \qquad \text{... (17b)}$$

Abbe's Sine Condition

It is a relation of great practical importance in Geometrical Optics and will be used later in deriving the magnifying power of a microscope.

Let QQ_1 be the image of an object PP_1 obtained by tracing the course of two known rays after refraction at a convex spherical surface separating a medium of refractive index μ_1 from a medium of refractive index μ_2. Let u, v, r denote the object distance, the image distance and the radius of curvature as usual.

An oblique ray PA making an angle α_1 with the axis meets the axis again at Q after refraction so that the refracted ray AQ makes an angle α_2 with the axis.

The line NAC it normal to the surface at A and let θ and ϕ be the angles of incidence and refraction respectively. As regards the signs of the angles involved refer to conventions (4) and (5). According to them α_1, θ and ϕ, are positive and α_2 is negative.

From similar triangles CPP_1 and CQQ_1, we have

$$\frac{QQ_1}{PP_1} = \frac{OC}{PC}$$

$$= \frac{MQ - MC}{MP + MC},$$

$$\text{or } conventionally - \frac{h_2}{h_1} = \frac{v - r}{-u + r},$$

$$\text{or} \quad \frac{h_2}{h_1} = \frac{v - r}{u - r}, \qquad \dots (a)$$

where h_2 and h_1 are the respective heights of the image and the object.

Applying the law of sines to $\Delta PA\ C$, we have

$$\frac{\sin APC}{r} = \frac{\sin PAC}{-u + r},$$

$$\text{or} \quad \frac{\sin \alpha_1}{r} = \frac{\sin(180 - \theta)}{-u + r} = \frac{\sin\theta}{-(u - r)} \qquad \dots (b)$$

Similarly, from ΔQAC

$$-\frac{\sin \alpha_2}{r} = -\frac{\sin \alpha_2}{r}. \quad \text{... (c)}$$

Dividing (*b*) by (*c*), we get

$$-\frac{\sin \alpha_1}{\sin \alpha_2} = \frac{\sin \theta}{\sin \phi} \times \frac{v-r}{-(u-r)},$$

or

$$\frac{\sin \alpha_1}{\sin \alpha_2} = \frac{\sin \theta}{\sin \phi} \times \frac{v-r}{u-r}$$

$$= \frac{\sin \theta}{\sin \phi} \times \frac{h_2}{h_1} \quad \text{... [by equation (a)]}$$

$$= \frac{\mu_2}{\mu_1} \cdot \frac{h_2}{h_1},$$

$$\left[\because \frac{\sin \theta}{\sin \phi} = \frac{\mu_2}{\mu_1} \text{ by Snell's law}\right]$$

whence $\mu_1 h_1 \sin \alpha_1 = \mu_2 h_2 \sin \alpha_2$. ... (18)

This relation is known as the sine condition. Since no where in deducing the result, the assumption of paraxial rays is introduced, the relation may be applied even if α_1 and α_2 are large, to any system of aplanatic surfaces, so that a point image is formed at *Q* for even wide-angled rays leaving *P* and passing through the system.

One important case of the application of this condition is that to the objective of a microscope.

The relation (18) can be written as

$$\frac{h_2}{h_1} \times \frac{\sin \alpha_2}{\sin \alpha_1} = \frac{\mu_1}{\mu_2}$$

Now, $\frac{h_2}{h_1}$ is called *linear transverse magnification* and $\frac{\sin \alpha_2}{\sin \alpha_1}$

represents *angular magnification.* So that, we have linear transverse magnification × angular magnification

$$= \frac{\mu_1}{\mu_2}.$$

For a single refracting surface of a small aperture, the sine condition becomes

$$\mu_1 h_1 \alpha_1 = \mu_2 h_2 \alpha_2 \qquad \text{... (18a)}$$

If there are $(n - 1)$ co-axial refracting surfaces, we have

$$\mu_1 h_1 \alpha_1 = \mu_2 h_2 \alpha_2 = \text{.....} = \mu_n h_n \alpha_n \qquad \text{... (18b)}$$

a relation often referred to as Helmholtz Lagrange's equation which may be stated as follows :

> "If a number of media are separated by spherical refracting surfaces with their centres on the same axis and are traversed by paraxial rays proceeding from a small object in one of the media then $\mu h \alpha$ is constant, where h denotes the size of the image formed in the medium of refractive index μ and α the angle which that part of a ray that lies in this medium makes with the axis".

Important note. *In using the general relations deduced above and in the subsequent chapters on lenses for solving numerical problems it is necessary to substitute the numerical values of the data provided subject to the convention of signs used in deriving the formulae.*

Example. *A small object is placed 50 cm from a convex refracting surface of radius of curvature 20 cm. If the surface separates air from glass of refractive index 1.50, find (a) the power of the surface and (b) the position of the image.*

(*a*)

$$P = \frac{\mu_2 - \mu_1}{r} = \frac{1.50 - 1}{20/100}$$

$$= +0.50 \times 5 = +2.5\text{D}$$

(b) $$\frac{\mu_2}{v} = \frac{\mu_1}{u} = \frac{\mu_2 - \mu_1}{r}.$$

Here, $$u = -50 \text{ cm},$$

$$\mu_2 = 1.50,$$

$$\mu_1 = 1.$$

$$\therefore \quad \frac{1.50}{v} - \frac{1}{-50} = \frac{1.50 - 1}{20},$$

or $$\frac{1.50}{v} + \frac{1}{50} = \frac{0.5}{20} = \frac{1}{40},$$

or $$\frac{1.50}{v} = \frac{1}{40} - \frac{1}{50}$$

$$= \frac{5-4}{200} = \frac{1}{200}.$$

$$\therefore \quad v = 1.50 \times 200 = 300 \text{ cm}$$

$$= \mathbf{3\ m.}$$

The positive sign indicates that the image lies on the positive side of the origin.

Example. *The incident face of a glass block is bounded by a concave surface of radius 3 cm. A small object 2 cm high is situated in air and on the axis 10 cm from the vertex. Find (a) the first and the second focal lengths, (b) the image distance, (c) the lateral magnificent and (d) the power of the surface. (R I. of glass = 150).*

(*a*) $$f_1 = \frac{\mu_1 r}{\mu_2 - \mu_1} = -\frac{1 \times (-3)}{1.50 - 1} = \frac{3}{0.50}$$

$$= \mathbf{6.0\ cm.}$$

$$f_2 = \frac{\mu_2 r}{\mu_2 - \mu_1} = \frac{1.50 \times (-3)}{1.50 - 1} = \frac{-4.5}{0.50}$$

$$= \mathbf{-9.0\ cm.}$$

(*b*) $$\frac{\mu_2}{v} - \frac{\mu_1}{u} = \frac{\mu_2 - \mu_1}{r}.$$

$$\therefore \quad \frac{1.50}{v} - \frac{1}{-10} = \frac{1.50 - 1}{-3}, \qquad [\because u = -10 \text{ cm}]$$

or $$\frac{1.50}{v} + \frac{1}{10} = \frac{0.50}{-3} = -\frac{1}{6}.$$

or $$\frac{1.50}{v} = -\frac{1}{6} - \frac{1}{10} = \frac{-5-3}{30} = -\frac{8}{30} = -\frac{4}{15}$$

(*c*) $$m = -\frac{1.50 \times 15}{4} = \mathbf{-5.625\,cm.}$$

$$m = \frac{\mu_1}{\mu_2} \times \frac{v}{u} = \frac{1}{1.50}\left(\frac{-5.625}{-10}\right) = \frac{5.625}{15}$$

$$= \mathbf{0.375.}$$

$\therefore$ size of image = 0.375 × size of object

= 0.375 × 2

= **0.75 cm**

(*d*) $$P = \frac{\mu_2 - \mu_1}{r} = \frac{1.50 - 1}{-3/100} = -\frac{0.50 \times 100}{3}$$

$$= \mathbf{-16.67\ D.}$$

Example. *A mark placed on the surface of a glass sphere is viewed through the glass from a portion directly opposite. If the diameter of the sphere is 10 cm and the refractive index of glass is 15, find the position of the image.*

The mark *P* on the surface of the glass sphere, is viewed through glass from the side *M*. The incident ray lies in the glass ($\mu_1 = 1.5$) and is refracted at the surface *AM* into air ($\mu_2 = 1$). Here, *conventionally* $u = -10$ cm ; $r = -5$ cm. Applying the relation

$$\frac{\mu_2}{v} - \frac{\mu_1}{u} = \frac{\mu_2 - \mu_1}{r}, \text{ we have}$$

$$\frac{1}{v} - \frac{1.5}{-10} = \frac{1-1.50}{-5} = \frac{-0.50}{-5} = \frac{1}{10},$$

or $$\frac{1}{v} = \frac{1}{10} - \frac{15}{100} = \frac{10-15}{100} = \frac{-5}{100},$$

$\therefore$ $$v = \mathbf{-20\ cm}$$

that is, the image is formed at a distance of 20 cm from the surface *AM* and is virtual.

Example. *An object is placed 50 cm from the surface of a glass sphere of radius 12 cm along a diameter. Where will the final image be formed by refraction at both surfaces ? Refractive index of glass = 1.60.*

For refraction at first surface with vertex *M*, let v' be the distance of the image from it, then

$$\frac{\mu_2}{v'} - \frac{\mu_1}{u} = \frac{\mu_2 - \mu_1}{r_1}.$$

Here $$\mu_1 = 1;\ \mu_2 = 1.60,$$

$$u = -\ 50 \text{ cm.}$$

$$r_1 = +\ 12 \text{ cm.}$$

$\therefore$ $$\frac{1.60}{v'} - \frac{1}{-50} = \frac{1.60-1}{12} = \frac{1}{20}$$

or $$\frac{1.60}{v'} = \frac{1}{20} - \frac{1}{50} = \frac{5-2}{100} = \frac{3}{100}$$

$\therefore$ $$v' = \frac{1.60 \times 100}{3} = \frac{160}{3} \text{ cm.}$$

This image serves as a virtual object for the second surface so that $(v' - 2r)$ becomes the object distance for the second surface.

For refraction at second surface,

$$\frac{\mu_2}{v} - \frac{\mu_1}{(v' - 2r)} = \frac{\mu_2 - \mu_1}{r_2}$$

Since $\mu_2 = 1;$

$\mu_1 = 1.60;$

$v' - 2r = (160/3) - 24 = +\ 88/3$ cm

and $r_2 = -\ 12$ cm.

$$\therefore \quad \frac{1}{v} - \frac{1.60}{\frac{88}{3}} = \frac{1 - 1.60}{-12} = \frac{1}{20}.$$

$$\text{or} \quad \frac{1}{v} - \frac{4.80}{88} = \frac{1}{20}$$

$$\text{or} \quad \frac{1}{v} = \frac{1}{20} + \frac{48}{880} = \frac{92}{880}$$

$$\text{v} = \frac{880}{92} = \mathbf{+9.5\ cm.}$$

The final image is real and 9.5 cm from M' on the side of the observer.

Bibliography

Addicott, F.T. : *Optics,* University of California Press, London, 1982.

Atherton, J.C. : *Manipulation of Flowering,* Butterworths, London, 1987.

Ball. N.G., M.B. Wilkins : *The Physiology of Plant Growth and Development,* McGraw Hill book, London, 1969.

Bemier, G., J.M. Kinet : *Physical Optics,* C.R.C. Press, Boca Raton, 1981.

Benjamin, W.A. : California. Boyer, P.D. : *The Enzymes,* Academic Press, New York, 1970.

Bull, H.B. : *An Introduction to Physical Optics,* F.A. Davis Co. Philadelphia, Eisenburg, 1964.

Canny, M.J. : *Phloem Translocation,* Cambridge University Press, New York, 1973.

Cohen, P. : *Control of Enzyme Activity,* Chapman and Hall, London, 1976.

Crafts, A.S. and C.E. Crisp, : *Phloem Transport in Plants,* W.H. Freeman, San Francisco, 1973.

Douglas, J.S. : *Advanced Guide to Optics,* Pelham Books, London, 1976.

Epstein, E. : *Geometrical Optics,* John Wiley, New York, 1972.

Epstein, E. : *Mineral Nutrition of Plants : Principles and Perspectives,* John Wiley and Sons, New York, 1972.

Esau, K. : *Plant Anatomy,* John Wiley and Sons, New York, 1965.

Fershl, A. : *Enzyme Structure and Mechanism,* W.H. Freeman, San Francisco, 1977.

Forward, D.F. : *Energy and Carbon Metabolism,* Academic Press, New York. 1983.

Ftzler M.E. : *Plant Lectins : Molecular and Biological Aspects,* Academic Press, New York, 1985.

Gauch, H.G. : *Inorganic Plant Nutritionm,* Dowden, Hutchinson and Ross, Pennsylvania, 1972.

Govindjee, : *Bioenergetics of Photosynthesis,* Academic Press, New York, 1975.

Green, J.R. : *An Introduction to Vegetable Physiology,* London, 1911.

Halevy, A.H. : *Handbook of Optics,* C.R.C. Press, Boca Raton, 1985.

Loewy, A.G. and Siekevitz, P. : *Cell Structure and Function,* Rinehart and Winston, New York, 1969.

Mazliak, P. : *Light and Optics,* Chapmann and Hall, London, 1973.

McNeil, M. Darvill : *Structure and Function of the Primary Cell Walls of Plants,* McGraw, London, 1984.

Novikoff, A.B. : *The Endoplasmic Reticulum : a Cytochemist's View,* New York, 1976.

Ogawa, Y. and R.W. King : *Geometrical Optics,* Springer-Verlag, Berlin, 1990.

Opik, H. : *The Respiration of Higher Plants,* Edward Arnold Publishers, London, 1980.

Rabinowitch, E. and Govindjee, : *Photosynthesis.* John Wiley, New York, 1969.

Rains, D.W. J. Bonner and J.E. Varner : *Light,* Academic Press, New York, 1976.

Rubin, C. and Rosen, O. *Protein phosphorylation,* 1975.

Ruhland, W. : *Encyclopaedia of Optics,* Springer-Verlag, Berlin, 1961.

Salisbury, F.B. and C.W. Ross, : *Physical Optics,* Woodworth Publishing Co., Belmont, California, 1967.

Siegelman, H.W. and G. Hind : *Optics and Light,* Plenum, New York, 1978.

Srivastava, H.S. : *Elements of Physical Optics,* Rastogi Publications, Meerut, 1990.

Thimann, K.V. : *Optics,* CRC Press, Florida, 1980.

Tolbers P.C. : *Enzyme Kinetics,* Chapman and Hall, London, 1977.

Tolbert, N.E. : *Microbodies—Peroxisomes and Glyoxysomes,* Oxford University Press, London, 1971.

Vince, Pruce, : *Light and the Flowering Process,* Academic Press, London, 1984.

Walsh, C. : *Geometrical and Physical Optics,* Edward Arnold, London, 1979.

Weinheim, Osbome, D.J. : *Optics and Light,* Elsevier Biomedical Press, Holland, 1978.

Zeevart, J.A.D. : *Light and Optics,* McGraw Hill, London, 1976.

Zimmerman, M.H. and J.A. Milbum : *Encyclopaedia of Optics,* Springer Verlag, Berlin, 1975.